Computer Programm

An Introduction for the Scientifically Inclined

Computer Programming

An Introduction for the Scientifically Inclined

Sander Stoks

The publisher offers discounts on this book when ordered in quantity for bulk purchases and special sales. For more information, please contact `sales@curly-brace.com`. There is also an e-book edition of this book available.

Visit Curly Brace Publishing on the web: `http://www.curly-brace.com`

Stoks, Sander.
Computer Programming: An Introduction for the Scientifically Inclined / Sander Stoks
Includes bibliographical references and index.
ISBN 978-90-812788-1-2
NUR code 989

ISBN 978-90-812788-1-2

Contents

Chapter 0
Introduction

Computers are useless. They can only give you answers.

Pablo Picasso

0.1 Intended Audience

This book is intended for first or second year university students of science (e.g., physics, mathematics, or chemistry), for use in an introductory computer programming course, or for high school students in their final year with an appetite for science.

0.2 Why Another Programming Book?

The use of computers is clearly a large aspect in many areas of science. Basic programming skills are part of virtually every science course today. However, these courses are still somewhat of a 'second class citizen' at many science faculties. They are often taught as an 'aside' by scientists emptying a bag of FORTRAN tricks onto an unsuspecting class of students, or are delegated to the local computer science faculty and are being taught 'the right way' but without a clear view of what these students will be using computers for.

Most 'beginner level' programming books do not focus on scientific use. Many of them seem to assume that an exercise *'implement complex numbers, along with some operations on them'* covers science enough and quickly move on to more interesting and useful stuff, such as a dozen different sorting algorithms (if they focus on computer science) and perhaps some check book balancing applications (if they are of the 'teach yourself programming in 24 hours' variety).

There also seems to be a common misunderstanding between computer scientists and 'other' scientists. Computer scientists seem to view the 'other' scientists as hopelessly old-fashioned, set in their ways, and wasting valuable computer time by running incomprehensible, spaghetti-like FORTRAN code. The other way around, to the 'other' scientists, the computer scientists sometimes come across as door-to-door vacuum cleaner salespeople, trying to convince them to ditch their old model (which still works perfectly!) in favor of the programming language-du-jour.

As it almost always does, the truth lies somewhere in the middle. The 'hard' sciences have a somewhat atypical use of computers, and it is my firm belief that many programming courses for science students fail to recognize that. On the other hand, computer scientists obviously know what they are talking about, and totally ignoring their insights and advice when teaching students the basics of programming is a bad idea.

So, in this book I will try to present the basics of computer programming in a way which satisfies the particular needs of science students, while at the same time trying to clarify some concepts from a software engineering and computer science point of view. Of course, there will probably be disagreement from both camps. If only over the programming language I chose for this book (C), which almost all computer scientists would argue is a terrible choice for an introduction to programming (preferring a 'cleaner' language like Pascal or even Java), whereas many 'hard' scientists won't even give anything but FORTRAN a second look. If the two approaches 'collide', I will choose the side of science. More on the reason for using C in §2.5.

This book does not teach object oriented programming. For 99% of scientific applications, 'old-fashioned', procedural programming is best suited for the job, and incurs the least 'overhead' in terms of learning curve and time-to-results.

It is also my firm belief that science *students* are somewhat atypical. As an anecdote, there is an 'educational' programming language in which the 'largest positive integer number still representable by the computer' is fixed at 10 000. This was most probably chosen to look like 'a reasonably big number' because the *real* largest number (which on most computers is something like 4 294 967 295) would look arbitrary and confusing to students. To a physics or mathematics student, however, the number 10 000 looks far more 'special' and confusing. After an explanation about how a computer stores integer values and how binary arithmetic works (which they probably already know), a simple explanation that—at least on a pure 32-bit system—the largest positive integer that can natively be represented is $2^{32} - 1$ (which is the number above) would immediately make sense. If these students were not interested in 'how things really work', they would not be studying science in the first place.

0.3 Assumed Prior Knowledge

No prior programming experience is required to read and understand this book. In fact, it is probably better to start from scratch than to try and get rid of 'bad habits' in an existing 'style'.[1]

Given the intended audience, a basic level of mathematics is assumed. In examples taken from science, I will not expand too much on the science behind them. Even if the subject of an example makes no direct sense to the student (because it comes from a different area of science), the basics of what needs to be solved should still make sense.

[1] It is interesting to note that nearly all programming books contain a similar paragraph, which would lead to the conclusion that all programming book authors feel that their own style is the only correct one.

0.4 How to Read This Book

In principle, all new concepts are presented 'in order', and each new chapter assumes that all materials covered in the previous chapters are mastered. Each chapter concludes with a synopsis of what was covered; if you are not entirely new to programming you may want to check the synopsis first to see whether a certain chapter covers anything you don't know yet.

The short sections on other programming languages can be skipped if all you're interested in is C (although I would not recommend you do this).

Finally, each chapter has some Questions and Exercises. Like learning a foreign language or studying a musical instrument, reading about it is not enough. You will need to practice. Only if you program regularly and 'get your hands dirty' will you be able to master programming. Some of the questions and exercises are more theoretical in nature, and you can skip those if you absolutely must. But I encourage you to try all the programming exercises, and every time you think 'what would happen if I changed this-or-that in the program?' – just try it out! You will notice that the first few programs you write will contain beginners' mistakes, and the sooner you leave that stage behind you, the sooner you can focus on the *purpose* of your program.

Some questions are marked with a star ($\star$) to indicate that they're a bit more difficult, or two stars if they're *quite* difficult.

Unless otherwise noted, all values are in SI units throughout the book.

0.5 Scope of This Book

There should be a separate course (like 'computational physics' or 'computational chemistry') for students interested to expand on this subject; this book only tries to provide a solid foundation to enter these courses. Although these courses may well be using a different programming language from C (most likely FORTRAN), it is expected that the fundamentals taught in this book will still be valuable.

This book will touch only lightly on specific algorithms used in computational science, and will only cover the basics. So, you will not find the Runge-Kutta method explained, but will find some general 'computer science' concepts like linked lists, even though they might not be immediately useful in scientific programming.

Lastly, there is no denying that many science students will end up in a programming-related job after graduation (if not before that). Having a decent foundation with some knowledge of software engineering concepts will give these students a head-start.

0.6 Acknowledgements

First of all, I would like to thank my family for their patience, since writing this book has been a multi-year project eating up far more time than anticipated. Furthermore,

I would like to thank the following people who provided valuable feedback on the manuscript in various stages of completion: Gert-Jan de Vos, Femkje Prandi, Alexander van 't Hooft, Geert Gerritsen, Olaf Veenstra, and Vincent Stoks. I am a stubborn person, so any remaining errors are my sole responsibility. I would also like to thank Judah Milgram and Joost Dierkse for contributions to the cover production. The fern leaf photo on the cover is by Manu Mohan.

You can visit `http://www.curly-brace.com` for up-to-date information regarding this book (including any errata).

Chapter 1
Computers

Where a calculator on the ENIAC is equipped with 18,000 vacuum tubes and weighs 30 tons, computers in the future may have only 1,000 vacuum tubes and perhaps weigh $1\frac{1}{2}$ tons.

Popular Mechanics, March 1949

1.1 Do You Need To Know?

Many programming teachers deliberately refrain from going into details about computer hardware. Their reasoning is that intimate knowledge of the way computers work actually hinders the development of a proper programming style. In fact, many prefer working with pencil and paper instead of even *using* a computer. A quote by Edsger W. Dijkstra, a famous Dutch computer scientist, is that "Computer Science is no more about computers than astronomy is about telescopes." This is why the term 'computer science' is rather misleading. Some people have therefore adopted the slightly different term 'computing science'.

But this book is not about computer science. Instead, it treats computers in the same way an astronomer might treat telescopes. It certainly makes sense for an astronomer to know how telescopes work, and understand the basics of optics. Computers to us are *tools*, and any craftsman will explain to you that a thorough understanding of your tools can be vitally important.

Another famous quote is that it shouldn't matter whether your programs are performed by computers or by Tibetan monks. While this is certainly true, 'we scientists' are not necessarily in the business of creating the most elegant programs that would thrill Tibetan monks reading them. Our job is to get some kind of calculation finished—preferably before our computation time on the faculty's super computer reaches our quota. When our methods to achieve that goal resemble fastening a screw with a hammer, so be it.

On the other hand, it is worthwhile to keep the Tibetan monk quote in mind. Because surprisingly often, the program that a monk would be happiest to perform, is also the best solution.

1.2 Hardware Overview

1.2.1 Processor

The 'brain' of a computer is formed by the *Central Processing Unit,* or CPU, also simply called 'the processor'. The CPU can actually only do surprisingly simple things, but it can do it very fast. What a CPU does is execute a list of instructions which it fetches from memory. These instructions form the so-called 'machine language'. They are the smallest 'building blocks' of a program. The atoms, if you will[1].

Examples of the kind of instructions a CPU can perform are 'retrieve the contents of memory location *p*', 'compare this value to zero', 'add these two values', 'store this value in memory location *p*', 'continue running the program from memory location *p*', etc. Everything more 'high-level' that a computer does, is expressed in lists of instructions of this very simple kind. Things like 'print this file', 'store these data on disk', 'check whether the user has pressed a key', or 'plot these data in a graph on screen' are totally alien to the CPU and need to be expressed in machine language programs that can be hundreds or thousands of instructions long. Don't worry though, you won't need to write those. We can use higher-level *programming languages* and *libraries* that provide an abstraction layer above the machine language level.

Machine language is usually referred to as *second generation programming language*, the first generation being the actual (binary) codes. The translation of more-or-less human-readable instructions to the actual machine codes is done using an *assembler*. The human-readable machine language instructions are called *mnemonics*, because they are easier to remember than the actual codes. So, the assembler might translate an instruction like `ADD A,(HL)` into 10000110[2].

A *third generation programming language*, which is the type we will be focusing on here, offers an even higher level of abstraction with named variables, English-like syntax, etc., which would let you write instructions like '`print y`' or '`result = (1 + epsilon)*sin(x)`'. Translation of your third-generation program into something the machine can handle is done by either a *compiler* or an *interpreter*—we'll get back to this later.

Research on *fourth generation programming languages* is in full swing; these would let you write 'meta-programs' like 'Write me a program to calculate the energy levels in a single quantum well with variable parameters.' Unfortunately, we're not quite there yet.

Typically, a CPU has in the order of a few dozen to about one hundred different instructions. The first processors did not even have instructions to, say, 'multiply these two values'; instead, this needed to be explicitly implemented in lower level

---

[1] Since you are an (aspiring) scientist, you may be wondering whether the equivalents of nuclear particles also exist, and yes they do. Internally, most modern processors actually have instructions on an even lower level, and the machine language is implemented on top of this.

[2] Although the following trivia is totally irrelevant for the present text, the instruction in fact means "Add the contents of the memory location at the address given by the register `HL` to the accumulator register `A`" in the machine code of the Zilog Z80 processor, as found in the Sinclair ZX80 and later computers (most notably the ZX Spectrum and the Tandy TRS80).

instructions (using repeated addition, for example). Although modern processors can have quite elaborate instructions that multiply vectors or compare two values and exchange them, the real power of processors is that they are *fast*. Typically, they can process millions of these instructions per second, with billions being no exception.

The speed of processors is often measured in Hertz. The value you often see quoted for a certain computer system is their *clock frequency*. One 'clock tick' is the smallest time slice a system can operate with. This is actually not a very good measure to compare different types of processors. The problem is that different types of instructions can take a different number of clock ticks to perform, so the clock speed is not trivially related to the number of instructions per second the processor can perform. There is a measure of processor performance called MIPS (million instructions per second) which for this reason is also known as 'meaningless indication of processor speed'.

1.2.2 Memory

Computers store, retrieve, and operate on data. These data are stored in some form of *memory*. To write efficient programs, it is important to understand that a computer has a form of memory *hierarchy*. There are different kinds of memory in a system, each with different properties.

The fastest kind of memory available to a computer are its *registers*. These are located in the processor itself, and can often be accessed in a single clock tick. Usually, the processor has relatively few of them available, in the order of only four to 128 or even 256 in more advanced processor architectures. Usually it is in the order of 32. The 'size' of these registers (determining the range of numbers they can store) is dependent on the architecture of the processor. For example, when a processor is said to be *32 bit*, the main registers of that processor are 32 bits in size, and it can usually address 2^{32} bytes of memory (hence 4 gigabytes is the upper limit of the memory size of a 32-bit machine).

The 'normal', or 'working' memory of a computer is called RAM, for 'Random Access Memory'. One can view memory as a (large) array of *locations* that can each store a *value*. The locations are numbered, and this number is called the *address* of the memory location. The smallest 'addressable' memory unit is the *byte*. Each byte consists of eight *bits* (binary digits).[3] These bits can have only two values: 1 or 0 (or 'on' or 'off', if you wish). Since a byte has 8 bits, it can store any integer value between 0 (all bits are 0) through $2^8 - 1 = 255$ (all bits are 1). See section 1.3 about binary arithmetic if this doesn't make sense to you.

To handle larger numbers, bytes can be grouped together to form a 16-bit *word* (storing up to $2^{16} - 1 = 65\,535$) or a 32 bit *long word* (storing up to $2^{32} - 1 = 4\,294\,967\,295$). On 64-bit systems, there also is a *long long word*, combining 8 bytes.

It is dangerous to mention typical memory sizes since that will make this book look hopelessly outdated in only a few years' time, but at the time of writing, memory

[3] Actually, this number needn't be 8. There have been computer systems with a 'native' word size of 7 bits, or even 6. You can quite safely assume that a byte is 8 bits, however, and if some wise guy wants to make you look stupid for making that assumption, ask him to show you a machine with a different number. A *working* one.

capacities of a few gigabytes[4] were becoming commonplace for personal computers. Multi-user systems in use at computer centers at universities can have (much) more, and capacities in the many gigabyte range are not unusual. 64-bit systems have been available for quite a while in 'scientific machines', and at the time of writing this book they were being adopted in desktop systems (and even laptops) as well. These systems can address more than 4 gigabytes of memory.

It is important to realize that RAM is relatively slow. Although memory technology progresses and memory is getting faster, processors have accelerated far quicker. Typically, retrieving the contents of a memory location in RAM takes in the order of 10 ns. Compare this to a clock tick cycle of 1 ns in a 1 GHz system. The processor would spend most of its time waiting for data to become available from memory.

To remedy this problem, *cache memory* was introduced. This is memory that acts as an 'intermediate'. It is fast (in the order of, say, 3 ns) but also far more expensive than 'normal' memory, and therefore a typical system has far less of it. The way it works is that it stores data as a buffer between the CPU and main memory. If the CPU asks for the contents of a certain memory location, the memory subsystem first checks to see whether it happens to be in the cache memory, and only retrieves it from main memory if not (and stores it in the cache too, overwriting 'older' data there). Since many computer programs look at the same data more than once, next time that data is requested a lot of time can be saved because it is still in the cache memory. Most computer systems have several levels of cache memory, usually 'level 1 cache' directly on the CPU, running at the same speed as the CPU itself (or, say, half of that), or 'level 2 cache' which is slightly slower (and cheaper, and thus larger). Typical values are to the order of 32 kilobytes of level 1 cache and a megabyte for level 2 cache. There is actually even more cleverness involved in the way cache memories work (for example, RAM often is more efficient if it is asked for the contents of several adjacent memory locations in one 'burst', so the memory subsystem could gather *more* data than it is asked for at the time and store it in the cache, based on the prediction that the CPU might need that data in the near future anyway). For more information, you can take a look at more specialized literature like [5].

Whereas cache operation is completely transparent to the programmer (*i.e.* you don't need to know it's there), programs can run much faster (up to an order of magnitude) if the program and the data it operates on 'fit' in the cache.

It is also quite important to recognize that the types of memory mentioned so far are *volatile*. That is to say, this memory only 'remembers' its contents while the system is switched on. Often, you would like to store data for a longer period of time. There are different types of memory which are *persistent*, such as a Indexhard disk, CD-ROM, or flash storage.

Typically, capacities of the persistent storage available to a computer system are far larger than the RAM size. Consumer-level hard disks have capacities in the terabyte-order. It also needs to be noted that this kind of storage is several orders of magnitude

[4]In line with the base-2 'nature' of computers, the SI prefixes are used in a (slightly) different meaning. A 'kilobyte' is not 1000 bytes but actually 1024 (2^{10}) bytes; similarly, a 'megabyte' is 1 048 576 (2^{20}) bytes, etc. The only exception is the size of your hard disk, which vendors usually express using the SI prefixes (because that yields larger numbers which look better on their spec sheet).

slower still than RAM. When you are operating on data sets that are really large (so they won't fit in RAM), this is something to take into account. For several areas of science (such as astronomy or experimental high-energy physics), huge data sets are the rule.

Hard disks come in two major flavors, IDE and SCSI. The former stands for 'Integrated Drive Electronics' and is traditionally common on personal computers, while the latter stands for 'Small Computer Systems Interface' and is the system of choice for multi-user or high-performance computers. Traditionally, IDE drives have been cheaper but slower, and SCSI has offered some nice advantages like being able to chain more devices together, and offer 'redundant storage systems' (*i.e.* store the same data multiple times, so that when one drive breaks down, the data is not lost). IDE has caught up quite nicely, and although the fastest and meanest disk drives are still SCSI, IDE suffices for most applications; especially with the advent of high-speed 'Serial ATA' (ATA is the 'official name' for what everyone calls IDE). Again, this difference probably does not need to concern you unless you are deciding on a hardware system for your specific experiment or calculations. If your application involves huge amounts of measurement data that need to be stored in real time, you should consider equipping your system with SCSI. For completeness, it should also be mentioned that SCSI is not limited to hard disks only, as there are other peripherals (like scanners) which connect to the system via SCSI, although this is superseded with USB and FireWire (see below).

1.2.3 Peripherals and Interfaces

Of course, there need to be ways to get information into a computer and ways to view results. Typically, a computer has several *input devices* like a keyboard or a mouse, and *output devices* like a monitor or a printer.

In experimental science, computers are often also used for controlling an experimental set-up, or for data-acquisition. For this, there are a variety of ways to interface ('talk') to the computer. For relatively slow connections, with data rates in the order of up to 10 kilobytes/second, it is often easy to use the *serial port*. 'Serial' means that the data bits are transferred one after another, as opposed to 'parallel', when multiple bits (mostly 8, or some multiple of 8) are transferred simultaneously. Obviously, a parallel port requires more physical wires. Traditionally, printers have been connected to computers using a parallel interface.

Most computers have several serial ports available which operate following the RS-232 standard. This is quite a popular interface because it is both well documented and relatively easy to implement with cheap off-the-shelf electronics, and runs at speeds up to 115 kilobits/second (230 or even 460 on some systems). As an example, most external modems work via this interface.

Incidentally, 'legacy' parallel and serial interfaces are slowly being replaced by more modern interfaces such as USB (see below). This has the benefit of not having lots of different interfaces which can each take at most one or two devices, having special interfaces for your keyboard, mouse, modem, etc. The drawback is that the modern interfaces are more complex and need integrated circuitry to connect, whereas the 'old' interfaces can easily be used in an experimental setup using an old-fashioned soldering iron and a simple wiring diagram.

On the other end of the spectrum are high-data-rate interfaces such as GPIB which require installing separate extension cards in the computer, driven by special software, but enabling far higher throughput. This is needed to do real-time readouts of oscilloscopes, for example, with sampling rates up to the order of 100 MHz or more.

In between of these extremes, there are interfaces such as USB (Universal Serial Bus) and FireWire (officially called IEEE 1394), which several more recent peripherals and/or measurement systems are supporting. These interfaces support 'hot-plugging', *i.e.* the peripherals can be connected and disconnected while the system is switched on, and are detected and configured 'on the fly'. There is a trend towards using ethernet as an interface to peripherals (especially more 'elaborate' devices); they often include a small embedded computer which is configurable via a 'web interface'.

It is important to note that often, measurement equipment produces data in *analog* form, which needs to be converted to the *digital* form which a computer can work with. Many data acquisition cards have analog-digital converters that can convert hundreds of millions of samples per second.

In a pinch, it is worth noting that a *very* cheap and easy-to-use digital-to-analog and analog-to-digital converter is present in almost any consumer-level personal computer in the form of its sound card. Whereas this is probably not suited for serious lab work due to the limited amount of sample frequencies it can work with and its resolution, it has been used successfully in a wide range of high-school or first-year science lab type experiments.

1.2.4 Networks and Clustering

To get more computing power, you could get a bigger computer, but you could also try to somehow connect multiple computers together in a network, forming a *cluster*. This approach doesn't always work; only when the specific calculations you need to do lend themselves to be split up in multiple, independent parts, which can then be calculated on separate computers. This is of no use when each of your calculations depends on the outcome of a previous one, since the computers in your cluster would spend their precious time waiting for another computer to finish its calculation, then do their part of the job, and finally hand off the result to the next. Also, there is considerable 'overhead' associated with splitting up the calculations. Network connections are slower than intra-computer connections, so sending lots of data back and forth is going to take a relatively long time. Only for large and time-consuming calculations involving relatively little data, it is worthwhile to use multiple computers.

Excellent examples of this 'distributed computing' on a very large scale are the 'SETI at home' project and the RC5 project. Both harness the 'spare computer cycles' of computers all over the Internet, using computers which their owners have registered with the project maintainers. The former project searches for extra-terrestrial intelligence by handing out packets of radio frequency readings to (home) computers, which then run a data analysis program on it when they would otherwise be sitting idle, and send the results back to the project maintainers (and then receive a new set, etc.). The latter project was more of a 'proof of concept', and was successfully used to crack a certain encryption scheme by brute force which would have taken tens of years in a conventional approach.

On a smaller scale, computer animation as used in films is often performed on clusters of computers called 'render farms', comprising hundreds of relatively low-cost computers and dramatically speeding up the rendering process.

Also, it is quite customary for computers to have more than one CPU, or have CPUs which internally have multiple complete processing units (so-called *multi-core* processors). This is an especially cost-effective way of increasing computer power, as prices increase exponentially with CPU speed (and CPU designers are running into physical limits regarding their clock speed) whereas performance increases only linearly. Performance does not scale exactly linearly with the number of CPUs though, since there is an overhead as well. Depending on the nature of the calculation and on how well the hardware and the operating system can cope with multiple CPUs, adding a second identical CPU will yield anywhere between 1.5 – 1.9 times the performance. Also, this extra performance does not come 'free': The programmer will have to make specific adjustments to the program to make it use the available CPUs.

1.3 Binary Arithmetic

Whereas most people use the decimal system, computers use a *binary* system. In this system, there are only two digits: 0 and 1. In our day-to-day decimal system, we are used to the fact that the *position* of a digit within a number determines its 'weight' when determining the value. The rightmost digit designates the 'ones', the one immediately to its left the 'tens', then the 'hundreds', etcetera. Formally, the value of a decimal number of n digits, numbered from right to left (!) as $d_n \cdots d_2 d_1 d_0$ is

$$\sum_{i=0}^{n} d_i 10^i.$$

So, the interpretation of the number 3207 in the decimal system is (going from right to left): 7 times 10^0 (7), plus 0 times 10^1 (0), plus 2 times 10^2 (200), plus 3 times 10^3 (3000), equals three thousand two hundred and seven. This is so trivial you don't usually stop to think about it.

Quite similarly, in the binary system, the rightmost digit designates the 'ones', the one immediately to its left the 'twos', then the 'fours', etc. So, for a binary number the formal interpretation would be

$$\sum_{i=0}^{n} d_i 2^i.$$

As an example, the interpretation of the binary number 1101 is (again, going from right to left): 1 time 2^0 (1), plus 0 times 2^1 (0), plus 1 time 2^2 (4), plus 1 time 2^3 (8), equals thirteen.

In the same way that powers of ten form 'natural orders of magnitude' for (most) humans, so are powers of *two* the 'natural orders of magnitude' for binary systems.

There is another system in regular use in computing, which is the *hexadecimal* system (often simply called 'hex'). This is a base-16 system, so it has a few *extra* digits besides our usual 0 – 9. In the hexadecimal system, these are designated using letters:

Hex	Decimal
A	10
B	11
C	12
D	13
E	14
F	15

Hence, the hexadecimal number 3E8B is interpreted as 'B' times 16^0 (*i.e.* 11 times 1), plus 8 times 16^1, plus 'E' times 16^2 (*i.e.* 14 times 256), plus 3 times 16^3, equals sixteen thousand and eleven (16011 in decimal). To differentiate hexadecimal from decimal numbers, they are often prefixed with '0x' (that's zero-x) or postfixed with 'H'. The above number would then be written as either 0x3E8B or 3E8BH. The hexadecimal system is not actually used by the computer itself, but rather in computer programming because the relation to binary values is clearer than when decimal numbers are used.

1.3.1 Negative Values

In our decimal system, we have a 'special digit' which can only occur at the leftmost position in a number, and which designates negative numbers. The previous sentence is, of course, just a convoluted description of the minus sign. In the binary system, there is no such special digit, and so far we have only seen how to represent positive integers (or zero) in binary. Clearly, there is a use for negative numbers, and there are several ways to represent them. The most often used system is called *two's complement*. It uses one bit (the leftmost) as a *sign bit*, using 0 for positive and 1 for negative. Also, to negate a number, each 1 in the binary representation is replaced with a 0 and vice versa, and finally 1 is added to the result. The main advantage of this 'agreement' is that you can subsequently perform arithmetic on these numbers without having to treat negative values in a 'special' way.

So, to negate the 8-bit binary value 0010 1101, we would first get 1101 0010, and then add 1 to the result, getting 1101 0011. Using this system, the largest negative value representable in 8 bits is -127 in decimal: 'minus' 0111 1111 becomes 1000 0001. The largest positive value is then 128 decimal; anything larger would have the 7th bit and another bit set, which would make it be interpreted as negative.

Therefore, it is important to realize that seeing only the digits of a certain binary number, say 1101 0011, doesn't tell you whether this number represents 211 or -45.

Incidentally, there is another system called *one's complement* in which a negative value is simply formed by inverting all the bits (*i.e.*, without adding one). This system has the drawback that there are two ways of representing 'zero', namely 'all bits cleared' but also 'all bits set'. The latter would correspond to 'minus zero'. This ambiguity has led to the adoption of two's complement instead in the majority of systems.

1.3.2 Floating Point Numbers

Apart from integer numbers of various sizes, computers can also work with *floating point numbers*, often simply called *floats*, or, in some computer languages, *reals*. It is quite important for scientific programming to realize that computer reals are not 'real' reals, in that they have a *finite* precision. More on that, and why it is important, in the next subsection.

The way a computer stores floating point values is rather clever because it allows a wide range of values to be stored in the same number of bits. Of course, to us scientists it is actually nothing new, as we often use a similar trick when dealing with either very large or very small numbers: We note a certain factor (with a certain precision) and add 'times ten to the power of n'. For floating point values, the computer simply divides the available space in two parts, and stores the factor (called the *mantissa*) in one part, and the exponent in the other. Both are signed with a single bit. Most systems offer two or even three varieties of floating point number types, with increasing numbers of bits available for increased accuracy, in a tradeoff for memory requirements and/or execution speed.

1.3.3 Range and Accuracy

For the integer types (bytes, words, and long words) it is quite obvious that there is a limit to the actual value they can store. Trying to store 70 000 in a 16-bit word simply won't fit. Nor will 40 000 in a 16-bit signed word. Trying to do so anyway will result in the computer signaling an error condition called *overflow*. What specific type of variable you will need to use in your programs to prevent this phenomenon looks to be simple enough at first sight. However, you must realize that this limit also has an effect on *intermediate results*.

As an example, consider you are writing a program to calculate the average distance to the sun for all the planets of our solar system. You decide to use integer variables (just for the sake of the argument), and since a quick glance at the distances table learns that the average probably comes out at about 2×10^9 km, you decide that it is safe to use 32-bit, unsigned integers (since these can hold over 4×10^9).

Now, depending on how you write your computer program, you might still run into trouble. If you would do it the 'naive' way, by simply adding up the various distances from the sun for each of the planets, and finally dividing by nine to get the average value, an overflow will occur during the calculation. This is because the *sum* of all these distances is close to 1.6×10^{10} km, and that intermediate result *doesn't fit*, no matter whether you are going to be dividing it into something more manageable later on.

Of course, the correct thing to do in this rather contrived example would be to use floating point values. They are called 'floating point' for a reason: when the value grows 'too big', the exponent changes to keep the mantissa within range. You can view it as if a float value 'resizes to fit'. For most systems, even the smallest type of floats can vary between 10^{-37} and 10^{+38}, and the mantissa has 24 digits of precision. If that is not enough, there is usually a 'double precision float' (or simply 'double') available, that often goes from 10^{-308} to 10^{+308} with 53 digits of precision.

However, it is important to realize that floats have a *finite* precision. For a human, the question 'how much is $10^{100} + 1$?' is just as easy to answer as 'how much is $10^{10} + 1$?' or 'how much is $10^{200} + 1$?'. For the computer, though, these pathological numbers pose a problem. Because they are big, the exponent shifts to make room (or 'the point floats', if you will), but adding that 1 will then be problematic because there is not enough precision left. This results in the somewhat disturbing conclusion that to a computer, $N + 1 = N$ for sufficiently large N.

It is quite important to keep these anomalies in mind when designing scientific computer programs.

1.3.4 Rational and Complex Numbers

Although some computer languages also have a special type of variable for storing complex numbers, many do not. This was added to the C standard relatively late (and not all compilers support it yet). This is one of the reasons scientists usually scoffed at computer scientists trying to sell them C over FORTRAN (which has had complex numbers for ages, as well as high-precision floats). We will see that it doesn't *really* matter as you can add your own types to most serious programming languages (including, in a limited and somewhat concocted way, C).

A more fundamental issue is that computers tend not to know about rational numbers. To a computer, $\frac{1}{2} = 0.5$, no matter how much the difference has been beaten into our heads at school (saying that $\frac{1}{2}$ is 'infinitely precise' and that 0.5 represents anything between 0.45 and $0.5499\cdots$). The computer cannot accurately represent $\frac{1}{3}$, for example. That means that there are fundamental rounding errors that could possibly cause trouble. This phenomenon is investigated later in this book.

Now, it needs to be said that with clever programming, one *can* make computers work with rational numbers (in fact, they can do algebra just fine). This has even appeared in the realm of handheld calculators. But at the lowest (hardware) level, the most advanced type of numbers a computer 'knows' about are floats (or possibly vectors of them, in more modern machines).

1.4 Operating Systems

The operating system your scientific program runs on is usually even less of an issue than the particular computer hardware. There is one notable exception which is in experiment control, which is why we will briefly dwell on the subject.

The first (big) programmable computers were quite literally 're-wired' for each new program. The operator would plug in cables, like an old-style telephone operator. Later, computers were re-programmable more easily by reading the programs from punch cards. These computers were operated in *batch mode*. That is to say, the programmer would write a program (usually in a low-level language at that time), get it punched in a stack of punch cards, and hand this to the computer operator. When it was this program's turn to run, the operator would load the program into the computer, run it, and when it was done, collect the output and run the next one.

This method of running computers turned out not to be the most efficient one, since when a program was loading a large data set off a (relatively slow) storage medium, the CPU would just sit there, twiddling its expensive thumbs waiting for that operation to finish.

With the advent of faster and larger memory, computers could hold several programs in memory at once, and 'switch' between them. *I.e.,* when one program would issue a 'slow' operation, the computer would pay attention to another program while, for example, the disk was loading the requested information into memory for the first program. Once that was done, the computer would switch back to the first one. This ensured that the CPU was always running at full throttle.

Running several programs 'at once' (note that it wasn't *really* 'at once', but rather 'small parts of them in rapid succession') introduced all kinds of other problems—for example, when one of the programs would have an error in it, say, overwriting the contents of random memory locations, other programs running along with it could be influenced by that, producing erroneous results, even though they were themselves fully correct.

Another problem is that of *limited resources*; suppose two programs request a chunk of memory for intermediate results 3/4 the size of the total memory. Were each of these running on the machine alone, this would not be a problem. But when running on the same machine simultaneously, the second program trying to get the requested chunk of memory would somehow have to be either told this failed, or suspended until the first program finished with it. The same goes for multiple processes requesting access to, say, a plotter or printer connected to the computer.

So, gradually the *operating system* expanded from a simple scheduler for various jobs, into a complex system taking care of memory management, resource allocation, etc. It also provides a variety of services to programs, such as 'abstracting' various types of computer hardware (so that your program does not need to know exactly what kind of sound card the computer system has, or what video card, etc.), and provides support for showing your program's output in a window on screen, which the user can move, resize, etc. It also takes care of file management on the computer's hard disks, provides network connections (to other machines on your local network or to the Internet), and much more. It is safe to say that for the vast majority of computers nowadays, the operating system itself is probably the most complex piece of software running on it.

Usually, an operating system tries to be as *transparent* as possible, *i.e.* programs would spend most of their time running as if they were the only program on the system, apart from some 'agreements' that a program never accesses hardware directly, but always 'asks' the operating system for it (to prevent the *limited resources* problem mentioned above), etc.

However, there are some situations in which it is important to realize that running on a system together with other programs imposes subtle differences with having the system all to yourself. Suppose the computer is being used to drive some kind of experimental set-up in which some apparatus is controlled, and some measurement data needs to be collected a certain amount of time after an event occurs. There are plenty of examples for such a set-up, such as firing a laser into a cell containing a gas mixture, and reading an image from a CCD camera exactly x milliseconds later.

Now, your computer program might in broad lines be structured like this:

1. Fire laser
2. Wait x milliseconds
3. Read data from camera

which looks easy enough. However, if the operating system decides to interrupt your program after it has just fired the laser, then turn its attention to several other jobs running on the same computer (for example, because you have moved the mouse, or because some network activity was detected and the computer needs to store incoming data somewhere, or whatever else might be going on on your computer), and only returns to your experiment-driving program a couple of milliseconds later, the data read in from the camera would be 'stale'.

Because of this problem, there exists a special class of operating systems called *real time operating systems*, which make certain *guarantees* about how long certain operations will take at most. In such an operating system, you could ask to wait *exactly* x milliseconds, and while the system would be free to do other stuff in that time span, it *guarantees* it will return to your program within that limit.

While this all may seem rather trivial, it is surprising how many experiments are driven using a computer running an operating system that is thoroughly unsuited for that task.

For an excellent text on operating systems, see [6].

1.5 Specialized Machines

It is worth mentioning that there are specialized computers that can do a certain type of operations very efficiently. For example, some computers have special arithmetic units that can perform calculations on whole vectors at a time. In many areas of science, vector (and matrix) calculations form an important part of daily life, and thus a machine which can speed up these calculations can save a lot of time. Another example is *parallelism*, *i.e.* having multiple CPUs running simultaneously.

Some of these 'extras' are handled transparently to the programmer by the operating system (for example, by scheduling different concurrent processes to various processors) or by the specific programming language used or the compiler used to translate it to machine code (for example, the compiler might recognize that you are doing vector arithmetic and could insert built-in machine code instructions for these). However, sometimes the programmer needs to specifically use these advanced features.

Although it is outside the scope of this book to dwell on these specialized subjects too long, it is worth noting that several of these features, which used to be strictly the domain of high-end computers, are finding their way to the user desktop as well. Machines with two or more processors in them are available off-the-shelf (although it is not unusual for high-end computers to have 64 processors or even more), and many processors have used ideas from vector-processing systems in their instruction set, usually under the name of 'multimedia extensions' or some-such.

1.6 Synopsis

Whereas designing your programs as machine-independent as possible will generally result in more elegant programs, blatantly ignoring the features and limitations of computers could result in inefficient programs, or even incorrect and unexpected results.

This chapter gave a brief overview of computer hardware and operating systems, and pointed out some pitfalls to avoid when designing scientific programs.

Also, computer arithmetic and native data types were explained, along with some caveats as to precision and range of them.

1.7 Questions and Exercises

1.1 Write your birth year in binary and hexadecimal notations.

1.2 Most processors have specialized circuitry for performing multiplications, and the specific values of the operands of a multiplication do not make much difference in the speed at which the operation is performed. However, 'older' computers could multiply by a factor of 2^n significantly faster than by a factor of, say, 3^n or 7^n. Can you explain why? Hint: can you multiply by a factor of 10^n faster than by a factor of 3^n or 7^n?

1.3 To get a feeling for data acquisition and the data rates involved, calculate how many bytes per second are transferred when capturing audio at CD quality (which is sampled at 44.1 kHz, 16 bits per sample, stereo). Could this be transferred over a serial connection as mentioned in §1.2.3, without any compression techniques?

Chapter 2
Programs

> The process of preparing programs for a digital computer is especially attractive, not only because it can be economically and scientifically rewarding, but also because it can be an esthetic experience much like composing poetry or music.
>
> *Donald E. Knuth*

2.1 Computer Recipes

To explain what a computer program is, invariably the metaphor of a recipe comes up. This metaphor is quite bearded, but since it also makes good sense, we'll add a few hairs to the beard.

At a low level, when a computer 'runs a program', it basically just executes a list of instructions, just like a cook might follow the instructions on a recipe. This metaphor describes only a certain kind of programs, since there are other, more modern 'paradigms' of programming. However, programs written using these other styles are ultimately converted to lists of instructions for the computer to execute, and a large part of scientific programming challenges can be adequately tackled using this paradigm.

So, let us take a look at a typical student recipe.

Noodles

Ingredients

- 1 plastic cup of dried noodles
- optional: 1 sachet of sauce (probably inside the plastic cup)
- water

Preparation

- Open plastic cup of dried noodles
- Remove sachet of sauce from the cup, if it's there

- Boil a sufficient amount of water
- Pour boiling water into the cup up to the designated fill line
- Close the plastic cup
- Wait 3 minutes
- Add the sauce
- Stir
- Wait another 2 minutes
- Stir again
- Add some more hot water if desired
- Enjoy! (Wine suggestion: Chateau Lafitte)

Now, although this recipe might appear very simple, there are quite a few things that can be remarked about this example. First and foremost, note that this recipe is a procedure to produce a desired output (a delicious and wholesome meal) from a specific set of ingredients, neatly listed at the top, via a well-defined set of steps.

Secondly, the recipe involves making *decisions* based on *tests*—even if they are not spelled out. One test, for example, would have been 'Prod in the noodles after the designated amount of time, and see whether they have the desired structure. If they don't feel right, add some more water' instead of just 'Add some more hot water if desired'.

Thirdly, and perhaps most importantly at this stage, notice how the recipe does not explain *everything*. For example, it just says 'Boil a sufficient amount of water', and not 'Place a sufficient amount of water in a kettle, put the kettle on the stove, take a matchbox, strike the match against the box until it lights, turn the gas knob, light the gas under the kettle, wait until the temperature of the water reaches 100° C.' It is assumed here that the chef knows how to boil water. This might seem a trivial observation, but it is a most important concept. A recipe explaining *everything* (imagine having to explain how to operate the tap to put water in the kettle—all the way down to describing the motor muscle control in the chef's hand when turning the knob on the tap) would quickly be far too long to be usable. This would be equivalent to the 'machine language' mentioned in the previous chapter.

Similarly, a recipe not explaining enough is of little use either. For example, while an experienced chef might find an instruction like 'serve with hollandaise sauce' totally adequate, a student cooking all by himself (or herself) for the very first time might be at a loss when the recipe says 'fry an egg'. If the student runs into something like 'serve with hollandaise sauce', there had better been a separate recipe for hollandaise sauce somewhere in the cookbook.

So, an important lesson here is that a recipe (and a computer program) is written with a certain 'assumed knowledge'. You don't explain everything every time; you explain it once, and then refer to it. Given recipes for noodles and for hollandaise sauce, you could easily combine these into a recipe for noodles with hollandaise sauce—preferably not by *copying* the respective recipes, but by saying something like

Noodles with Hollandaise Sauce

- Prepare noodles
- Prepare hollandaise sauce
- Put noodles in a bowl
- Add hollandaise sauce
- Mix thoroughly
- Enjoy!

The 'Prepare noodles' and 'Prepare hollandaise sauce' are instructions of a higher level than the others, because they refer to other recipes. A similar thing is very common in computer programs. Even if specific 'sub-tasks' are only used in one place in a program, it can be quite useful to 'break them out' into separate parts of the program, because this makes the core of your program much more readable and gives a nice overview of what the program is supposed to do.

2.2 A Real Program

Enough with the metaphors; without further ado, let's proceed with a 'real' program. The following program lets you type two numbers and prints out the sum. It is written in BASIC, which stands for Beginners' All-purpose Symbolic Instruction Code. BASIC is one of the earlier programming languages, and is not considered suitable for any 'real work'. However, its syntax is closer to 'plain English' than that of C, and we'll compare the equivalent program in C in a while.

```
PRINT "Please enter two numbers"
INPUT a
INPUT b
LET c = a + b
PRINT "Their sum is "; c
END
```

It is almost possible to read this as a recipe written in plain English. Even if this is the very first time you've seen a program, you can probably guess what it does. We'll walk through the program explaining what everything does in detail below.

What you see here is a list of 'statements' (or 'commands'), one per line. In older BASIC dialects, every line had a line number. In more modern ones, this is not necessary. The line number also functioned as a 'label' and you could change the 'flow' of your program by using a `GOTO` *line number* command, after which the program execution would continue at the specified line. The use of `GOTO` is religiously frowned upon by computer scientists, because it facilitates a programming style which is affectionately known as 'spaghetti code'.

The `PRINT` command simply prints what comes after it to the screen. In this case, the text after it is surrounded by double quotes to denote that it is a *text string*, *i.e.* a piece of text that should be printed verbatim to the screen without further interpretation by the computer.

Next come two lines with `INPUT`. As you have probably guessed, these ask you to enter something into the computer. The little `a` and `b` are *variables* – just what you'd expect them to be. So, the numbers you type to the computer (closed with Enter or Return, or whatever your computer keyboard calls it), are stored in the variables named `a` and `b`. Variables are of the floating point type by default in BASIC. How the computer denotes that you are expected to enter something is implementation-specific; on some systems and dialects of BASIC, it might print out a question mark; on others, it might just flash a blinking cursor at you.

Next comes a line containing the `LET` command. What it does is assign the value of the expression `a + b` to the new variable named `c`. In most dialects of BASIC, you can actually leave out the `LET` command, so the line would just read `c = a + b`.

Then, the value of `c` is printed to the screen. You see that the text string '`Their sum is`' is followed by a semicolon, which in most BASIC dialects means 'more data follows, and you should print the rest on the screen immediately following what you just printed'. Had it been a comma, there would have been a 'tab' spacing on screen. Forget about these details though. What's important is that, since `c` is not surrounded by quotes, it is interpreted to be a variable, not a text string, and its value is printed to the screen.

Finally, the last line says that the program ends here. This particular statement is also not always needed – most BASIC dialects understand that the end of the program is denoted by the end of the text.

To actually execute this program, you'd have to find a computer with a BASIC interpreter, key it in, and type `RUN`. Yesteryear's computers of the 'home computer' crop, like the Sinclair ZX Spectrum or the Commodore 64, had a BASIC interpreter built in, so immediately after switching on the system, you would be able to key in our little program (each line prefixed with a line number, though), and run it.

2.3 The Same Program in C

At first glance our little program, now written in the computer language C (which we will use for the remainder of our book), will perhaps look rather intimidating compared to the friendly BASIC. Don't worry; although we won't do a similar walk-through as we did in the previous section, you will shortly be able to understand the details. For now, if you can vaguely recognize the main characteristics, you'll be fine.

```
#include <stdio.h>

int main(void)
{
    float a, b, c;
    printf("Please enter two numbers: ");
    scanf("%f %f", &a, &b);
```

```
        c = a + b;
        printf("Their sum is %f\n", c);
        return 0;
    }
```

The first thing you have probably noticed is that the C program looks more 'complicated' than the equivalent BASIC one. There are more strange characters (percent signs, number signs, curly braces, and ampersands), and every statement ends in a semicolon (this means you are free to put multiple statements on the same line, each terminated with a semicolon, but that is considered bad programming style because it makes your programs harder to read).

This 'syntactical complexity' is one of the main complaints many people have against C as a first language, but 'power comes at a price' and once you have a firm grasp of the language, you can write compact and powerful code. Also, many languages developed after C have 'borrowed' this syntax style, including the curly braces to delimit 'scopes' (more on that later), the semicolon, the function call syntax with a comma-separated list of arguments between brackets, etc. Knowing C, you can read programs in a variety of other programming languages as well.

2.4 Running Your Programs

By now, you are probably interested in actually running a program on a computer. Unfortunately, the precise steps required to do this vary quite a bit from one system to another. You will need a computer system with a C compiler installed (if you don't know what a compiler is, don't worry: you will two paragraphs from now). It falls outside of the scope of this book to expand on various compiler flavors, and we will limit our treatment of C as much as possible to ANSI C, which should run on any system.

Since you are probably interested in what goes on exactly when you make your computer run one of your programs, we will briefly go over the various steps here. This is necessary to get your first experimental C program running on your system, and some of the details will become more important later. Once these steps are defined, we will finally type up our first little C program, compile it, and run it. That's a promise.

Since the term *program* is used to mean both the code you typed and the final, runnable program, we will introduce some precise terminology here to eliminate confusion. By *source code*, we mean the program 'text' as you type it. This doesn't mean much to the computer. It's just a text file, and it cannot be executed. Instead, you use a special program called a *compiler*, which translates your source code to *object code*. Finally, a second program called a *linker* massages your object code, along with some 'glue', into an *executable* which can run on your computer system. The final product is also called a *binary*. The entire process is sometimes called *building* a program. It is worth noting that your source code is (strictly speaking) machine-independent, whereas your object code is tied to the specific machine you compiled it for.

In general, the process to build a program can be visualized with the following figure:

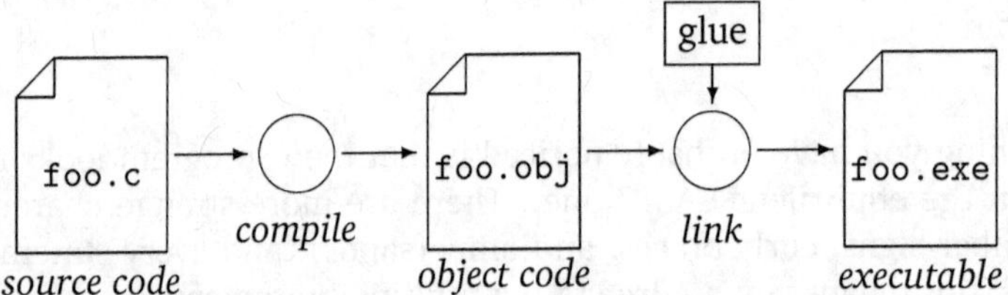

The file extensions in this figure have been chosen from the Windows system. On UNIX systems, the default extension for object files is .o, and executables usually have no extension at all.

Basically, getting a running program requires the following steps:

1. Type in your source code
2. Compile it to object code
3. Repeat the above steps until the compiler can't find any more errors
4. Link the object code to form an executable program
5. Execute the program.

For now, you can view the compile and link steps as one. For programs consisting of a single file (the majority of our programs, and in fact quite a large number of programs as used for scientific calculations), the compile and link steps are actually done with a single command.

Let us now finally type in our first program. The precise steps needed to compile and run it will be specified below for both a Microsoft Windows system and a generic UNIX system. So, open up your favorite text editor, and type the code below. Be wary of typos, because most will cause your compiler to 'emit scary warnings'. Also, note that I said 'text editor' and not 'word processor'. It is important to understand that a C program is just a regular text file, and there's nothing 'magical' about it. On the other hand, if you use a word processor (such as Microsoft Word), usually there are 'hidden' control codes embedded in your document (for instance, to specify the layout of the text, whether a certain phrase is in italics, etc.) and these would confuse the compiler. On Windows, you could use Notepad, for instance (crude though it is); if you're using a UNIX system you probably have a favorite plain-text editor already. There are many programming-oriented editors on either platform, and many good ones are even free. A quick Google for 'programming editor' or something like that should give quite a few hits.

```
#include <stdio.h>

int main(void)
{
    printf("Hello, world!\n");
```

```
    return 0;
}
```

After you've made sure you typed in your program exactly as shown above, including the weird # character and the semicolons, save this file under the name `hello.c`. It is a general convention to designate C programs by appending a `.c` extension to the file name, and there are operating systems which get hopelessly confused if you do not comply, so it is best to stick to it.

Don't worry if you don't understand all the details of the program. The occurrence of a `\n` character at the end of your greeting, and the curly braces, and the `int`, `void`, and `return` stuff – it is all explained in the next chapter. This is only to get something happening on your computer, and to get the pesky details about building programs on your particular systems out of the way.

Now, you have entered your program into your computer, but you cannot run it. You can email it to some Tibetan monks to admire, or compile it for execution on your own computer. We will assume you will want to do the latter. Depending on whether you are working on a UNIX system or on a personal computer with Windows, you can find below how to compile (and link) this program.

2.4.1 Building on Windows

If you have a computer with Microsoft Windows and have Developer Studio (MS Visual C++) installed, there is a compiler available under the name `cl` ('cee ell', not 'cee one'). This is a C++ compiler, but it also compiles 'plain' C. To be able to run this compiler from the command line (*i.e.* from a DOS prompt), you need to set some paths correctly. This is easily done by running the `VCVARS32.BAT` batch file, which comes with MSVC. It can probably be found at a location like `\Program Files\Microsoft Visual Studio 8\VC\bin\VCVARS32.BAT`, but the actual location may vary. If you have installed MSVC yourself, you are probably more than smart enough to find it. If you are working on a system that is remotely administered, there are probably some people in your organization who get paid to help you find it. Depending on how Developer Studio was installed, there may even be a handy shortcut in the Start menu to launch a command prompt window with the paths already set up correctly.

Once you've run this batch file in a command prompt window, you can then run `cl hello.c` (assuming you are in the directory containing your `hello.c` source code), which will invoke the compiler on your little program. It will perform compiling and linking in one go, and generate `hello.exe` in the same directory, which is the executable program. In the process, the compiler will generate a `hello.obj` file, which is the object file. You can safely throw this away; it is re-built every time you compile your program again.

If you are using MSVC, you can switch on the maximum warning level using the `/W4` switch:

```
cl /W4 hello.c
```

This is a good habit, and you should always compile at the maximum warning level. In §2.6 we will talk about these warnings.

By the way, if you indeed have Developer Studio installed, there are easier ways to edit, build, and run your program. Developer Studio is actually what is called an 'Integrated Development Environment' (IDE), and you can invoke the compiler with a keyboard shortcut. The editor does 'syntax highlighting' (it will give keywords it recognizes a certain color, which is a nice aid to catch typing errors while editing, since if a keyword in your program is spelled wrong, it will not get the color you expect). Also, it helps you find errors by jumping directly to the line in which the error occurred.

If you're using Developer Studio, you can use the 'project wizard' to set up a new project. For all the examples in this book, you should pick 'empty console project', which will set up a new project without filling in any code. You can also select 'Hello World project', but that would be cheating. If you select 'Run' from the appropriate menu, it will automatically re-build the executable if you have edited the code since the last time it was built. We'll get back to features like this in chapter 10.

Note that there are several other C compilers available for Windows, which vary wildly in price, performance, and quality. Some have different licensing options for non-commercial use, or offer special academic discounts. Examples are the Digital Mars compiler, the Intel compiler, and the Borland compiler. A quick Google for 'free Windows C compiler' may turn up some interesting results.

2.4.2 Building on UNIX Systems

Most vendors of UNIX systems ship a compiler with their system. Usually, you can invoke it with the `cc` command. On some systems, typing `cc hello.c` will actually only compile, but on most, it will also link the executable for you as a convenience. If not, you will have to link it yourself. In this case, you will find that the compiler generated a `hello.o` file, which is the object code for your program. To link this into an executable, you will need to invoke the linker. On some systems, this is also done with the `cc` command; often you can specify the name of the resulting executable with a `-o` switch, so you would type `cc -o hello hello.o`. On other systems, the linker has to be invoked with `ld`. You will need to do some experimenting to find out what the exact incantation for your particular system is. Sometimes, the name of the resulting executable will be `a.out` (for historical reasons) if you do not specify a name yourself.

Incidentally, there are actually some freely available UNIX-like operating systems out there such as Linux or FreeBSD. These often come with GCC, for 'Gnu Compiler Collection'. This includes an excellent C compiler. GCC is also available for many other operating systems, such as BeOS, and can even be installed on a Windows machine (with some extra effort – Google for 'MinGW').

With Mac OS X, you have access to a free-of-cost development environment as well (called Xcode), which is based on GCC.

If you are using GCC, it is a good habit to specify the `-Wall` flag, which turns on all warnings (we'll get to these warnings in §2.6). So the example above would be

```
cc -Wall -o hello hello.c
```

In chapter 10, we will take a look at some more advanced software building tools, in case your computer programs span several files. For the most part of this book, programs will usually be small enough to fit in one single `.c` file.

2.4.3 Running Your Hello World

To execute your first masterpiece of programming, simply type its name at your command prompt (you may have to prefix it with `./` if you're on a UNIX system). Your computer will run your program, and the text 'Hello, world!' should appear in all its glory.

While you are still all proud and warm and fuzzy, let us take a short detour and look at programming languages in a bit more detail. After that, we will take a look at what would have happened if you were unfortunate enough to make a typing error in your program. This would most likely have caused the compiler to complain with an error, which we will examine in the section after the next.

2.5 Programming Languages

We have now seen a similar program in two programming languages, BASIC and C. There are *lots* of programming languages, each with their own pros and cons. In general, you can divide programming languages in two broad categories; languages which are *compiled* and languages which are *interpreted*.

During the above paragraphs on how to build and run your program, you may have observed that there are quite a number of steps to take before you can actually run your program. Especially when you are making small changes to your code to see what effect they have on the program, it is a nuisance to continually have this cycle of 'edit, save, compile, link, run', and it would be nice to have a more 'interactive' way. When working like this, interpreted languages are nice. They don't need to be compiled and linked; instead, there is a 'runtime' which reads the program source and executes it line-by-line as it goes along. Examples of programming languages that work this way are BASIC and most scripting languages (most notably javascript). The latter is often used in small web-based applications since it can be embedded in a web page. In this case, the 'runtime' is part of your Internet browser.

There are two major drawbacks to interpreted languages. One is speed; you always have the overhead of the interpreter itself. For this reason alone, nobody uses interpreted languages for any significant scientific work. The other is that a compiler actually provides a first 'sanity check' on your code. If you make a typing error, for example typing `prinft` where you mean `printf`, the compiler will catch that (see next section). In an interpreted language, it will only be found at run time. In a program with considerable complexity, the typo may be in a line which gets executed rarely – meaning your program can be running fine for days, and suddenly stop. Also, for the same reason, it is hard for interpreted languages to implement *type-safety* (more on that later).

The most common languages compiled to machine code are Pascal, C, C++, and FORTRAN. Traditionally, the former is used in programming courses because it is clean, strict, and verbose, leading to structured and well-behaved programs. FORTRAN is the language of choice for many scientific programmers because it handles complex numbers natively and treats matrices and vectors as 'first class citizens', has millions of lines of code available in scientific libraries for many applications, and because the raw performance of its compiled code has always been the mark to beat for other languages. Unfortunately, it is also diagonally opposed to Pascal in imposing 'arbitrary' constraints on the way you write your code (actually, *because* of some of these constraints it can optimize so aggressively), leading to code which is hard to understand and maintain. With newer FORTRAN 'dialects' like FORTRAN90, many of these drawbacks are no longer there. Unfortunately, FORTRAN compilers are not nearly as ubiquitous as, say, C compilers.

There are also languages which are a bit of both. Most notably Java (which, despite the name, has nothing in common with javascript) and C# (pronounced 'C-sharp') are compiled, yet not to native machine language but to an 'intermediate' code which runs on a 'virtual machine'. Apart from only having to compile your programs once and being able to run this 'pseudo' machine code on multiple platforms (as long as there is a virtual machine available for that particular platform), it has the added advantage of only allowing a subset of what one would allow a 'real' program. For instance, one can deny such virtual machines any write-access to your hard disk. These kind of programs are often used as 'applets' you can download via the web.

C was traditionally intended for 'systems programming'; for writing, say, an operating system. It originally lacked scientific niceties such as the built-in complex type of FORTRAN, and it allows far more 'direct to the metal' programming than Pascal. This sounds like C is 'worst of both worlds'. The nice thing about C though is that you can write programs which are (nearly) as performant as FORTRAN programs, you can see 'what's going on' in terms of how the computer will actually do things (which would be considered a drawback to most 'purists' but without which a typical science student would be left with nagging questions), and that C compilers are everywhere. *Every* serious computing platform has a C compiler available, and C has become the 'lingua franca' of computers. Some would say this places it in the same category as Latin or Ancient Greek: Interesting for historical reasons, but not used anymore 'in the real world'. On the other hand, mastering these languages can be a great help when learning modern languages. The same goes for C: many more recently designed computer languages borrow heavily from C. Also, C continues to be used for many current projects, among which are some of the most high-profile computer programs in the world (such as the Linux kernel and many applications running on it).

Lastly, C++ is an 'extended version' of C which allows for *object oriented programming* and *generic programming*, both of which are outside the scope of this book. For most scientific type of programs, neither way of programming adds many benefits. A nice thing to note is that any valid C program is also a valid C++ program.

At certain points in this book, we will compare how other languages handle the concepts just introduced. If you are not interested in other programming languages or you are afraid you will become confused, you can skip these sections without missing anything important.

2.5.1 Variations in C

C is not the youngest of languages, and since its inception has undergone a few changes. The first major version of C was marked by the publishing of a book (in 1978) by its inventors, Brian Kernighan and Dennis Ritchie. After their names, this version of C is called 'K&R C'. There was no formal standardization of the language, and various implementations interpreted the book in slightly different ways. When the popularity of C grew (borrowing back features from C++ in the process), it became clear that a proper standard was needed, and the resulting version is usually referred to as 'ANSI C' or 'C89'. Kernighan and Ritchie updated their book to reflect this new standard ([2]). This book is highly recommended and can usually be obtained rather cheap, since book stores figure that a computer book this old can't be worth much anymore.

In the late 1990s, the C standard was revised again by an ISO committee, and the resulting standard is usually called 'C99' (the official document is ISO/IEC 9899:1999) because the standard was ratified by ISO in 1999.

Most compilers support C99, but not all compilers support it entirely. Therefore, the examples in this book will be based on C89 as much as possible, with a note explaining if anything changed in C99.

2.6 Errors and Warnings

Computer programs need to conform to a rather strict syntax for the compiler to understand them. It is important to note that the compiler can not catch any *semantic* errors. There are even compilers who will say 'None of the errors found' when successfully compiling a program. To understand the difference more closely, take a look at the following programs.

```
#include <stdio.h>

int main(void)
{
    float a, b, c;
    a = 10;
    b = 3.1415;
    c = a + b;
    printf("The sum of a and b is %f\n", b);
    return 0;
}
```

Even if you can't follow all the details (which is understandable, especially for the `printf` statement), you should see that it prints out 'The sum of a and b is' followed by the value of `b`, instead of `c`. This program, however, is perfectly legal C and no compiler will complain. The compiler cannot know that this program prints out something silly. Semantic errors like this are also sometimes called *bugs*: the program is 'correct' according to the compiler, but doesn't produce the intended result.

Now, if you have made a *syntax* error, the compiler will be able to produce an error message, usually telling you what it was expecting, and where in your source code the error occurred. The format of these error messages is not standardized, and they vary wildly between compilers.

For example, take the following program:

```
int main(void)
{
    float a, b, c
    a = 10;
    b = 20;
    c = a + b;
    return 0;
}
```

Apart from the fact that this program doesn't really do anything useful (since it doesn't print out the result of the calculation), note that there is no semicolon at the end of the first line. In C, all statements must end with one, so this is a syntactical error. Let's try to compile it anyway (forgetting semicolons is one of the things beginning C programmers do all the time, so it's better to get used to it).

Trying to compile it might give an error message like the following:

```
error.c: In function 'main':
error.c:4: parse error before 'a'
```

Note that it mentions the file it was working on (in this case, we assumed that the source code was in a file named `error.c`), what function it was working on (we'll get to this in the next chapter), and even what line it was on (line 4) when it encountered a 'parse error'. Your actual compiler output may vary, but it is typical that it didn't say something like 'missing semicolon on line 3'. In C, statements may span multiple lines, so for all the compiler knew, everything was fine and dandy at line 3, and only when it suddenly found that 'a' at the beginning of line 4, it suspected something was wrong. So a first tip when getting rid of syntax errors is to check out the preceding line(s) of your program.

Also, it is worth noting that a single mistyped line can mess up the compiler's understanding of all the code below it, in a sort of 'domino effect'. So, don't get scared if you get an enormous list of errors the first time you compile your program. It may well be that they all disappear once you fix the first one.

Apart from errors, the compiler can also emit warnings. In case of a warning, the compiler can continue compiling your program, but found some code which it found 'suspicious'. Most compilers let you specify a warning level (where a higher level means that it will be more verbose in venting its opinion about your code). *Always* compile at the highest warning level, and never ignore the compiler's warnings. After all, it knows quite a bit more about C than you do, and if it detects something fishy, you'd better take a second look.

Different compilers generate different warnings. Some people even use several different compilers on their code, each at maximum warning level, and only rest once none of these compilers finds anything to warn about.

Now that we have basic knowledge about building and running programs, let us cover the fundamentals of the C programming language in more detail. This will be the subject of the following chapter.

2.7 Synopsis

This chapter gave an introduction to computer programs, and how to compile and build them into executable binaries on different computer systems. You have met the 'hello world' program, saw an overview of various programming languages, and had a brief introduction to compiler errors and the difference between syntactic and semantic errors.

As many other languages, C is what is called a *compiled language*. A program written in C cannot be directly executed by the computer. It must be *compiled* (translated from *source code* into *object code*) and *linked* (combined with some 'glue' and made into a proper executable for this particular computer system) before it can be run.

2.8 Questions and Exercises

2.1 Insert some deliberate syntax errors in a program and try to compile it. Examine the error messages your C compiler gives you. Can you easily find the location of the error in your source file?

2.2 Would it be feasible to try and build a compiler which could detect semantical errors?

Synopsis

Questions and Exercises

Chapter 3
Functions and Variables

Home computers are being called upon to perform many new functions, including the consumption of homework formerly eaten by the dog.

Doug Larson

3.1 Hello World, Again

Let us go back to our first little 'Hello, World!' program and go over it in detail.

```
#include <stdio.h>

int main(void)
{
    printf("Hello, World!\n");
    return 0;
}
```

The C programming language actually has very few 'reserved words', *i.e.* words that mean something to the compiler straight away. In fact, there are three in the above program, and those are the words `int`, `void`, and `return`. We'll see the full list of reserved words in §3.7. The word `main` is also 'somewhat' special, but we'll get to that later.

Remember how we compared computer programs to recipes, and observed how we could make a recipe for noodles in hollandaise sauce by combining the recipe for noodles with the one for hollandaise sauce, and *referring* to them instead of explaining them in detail.

Our little program does a similar thing. The `printf` statement we use to print something to the screen is not a part of C, in its strictest sense. Instead, it is defined in terms of more basic, detailed code elsewhere. C comes with a large collection of such 'pre-defined recipes', which you can use as a programmer. The only thing you need to do, is explicitly tell the compiler what you assume to be 'prior knowledge' for this particular program.

This brings us to the first line in our program, which reads #include <stdio.h>. This is C's way of saying 'For this recipe, I assume you know how to make hollandaise sauce', or in this particular case, 'I assume you know how to printf'.

We'll leave it at that for now, but you will soon gain a deeper understanding of how this works 'behind the scenes'. For example, you might wonder whether #include is a special C keyword. Strictly speaking, it is not, but we'll get to that shortly.

The 'pre-defined recipes' are organized in *libraries*. For example, there's a bunch of code relating to input and output (e.g. getting data into the computer, writing stuff to files, and presenting it to a user) located in the library stdio – the 'standard I/O (input/output) library'. stdio.h is actually the name of a special file, namely the *header file* for this library, and it contains all the definitions so that your compiler will 'assume everything in that library as standard knowledge', so to speak. We will look at libraries in depth in chapter 12.

Now as to why #include is not a part of C per se. Any line in your program starting with # is considered a directive for a special program called the *preprocessor*, which parses your program before handing it off to the real compiler. Usually, the preprocessor is tightly coupled to the compiler itself (although you can run it separately, should you wish to do so). What happens when the preprocessor sees an #include directive is that it finds the file named between the <>-brackets, and inserts it verbatim as if you typed it right there in your program. To the compiler, it looks like you did.

Technically, you don't *need* to do this. If you leave it out, the compiler will issue a *warning* (only if you tell it you're interested in warnings, which you should) telling you it came across this new printf thing it doesn't know about, but it will continue regardless. Later, the linker will see the printf, look it up in the library, and link the corresponding code into your application (hence the name *linker*). In this case, that'll happen automatically. In case you make use of functionality from a more esoteric library, you will have to tell the linker where to look. Don't worry about this yet though.

Now, some more about that printf-function. The f stands for 'formatted'. It prints the string of characters to the screen (well, actually to 'the standard output device' but for now, just read 'the screen'). A 'string of characters' in C is designated with double quotes "like this". The \n means 'newline'. The backslash (\) is the 'escape character' in C, meaning 'interpret the character immediately following in a special way'. There are other 'escape sequences' as well, such as \t for a tabulation character and \b for 'backspace'. If you want to get a 'real' backslash (in some operating systems, it is used as a directory separator for example), you can type \\. You may remember that the percent sign means something special to printf as well, and you are right. We'll get to that once we have a more solid understanding about variables and types.

3.2 Variables and Types

Any C program is a collection of functions and variables. Every variable in C has a *type*. The same applies in day-to-day mathematics, but it is rarely explicitly mentioned.

More likely, the type of a variable is hinted at by the context or the name (most scientists use things like n for integer numbers (or N, if they're big), and x and y for real numbers). Anything with an i in the immediate vicinity is probably complex. And everything coming out of a sine function is a real number, etc.

In C, the type of a variable is quite explicit. If you want to have a variable `a` that is supposed to represent some real number, you must state in your program:

```
float a;
```

after which `a` will 'forever' mean 'a variable of floating point type' (for a well-defined meaning of the word 'forever', but we will get to that). Such a statement is called a *declaration*: 'I hereby declare that henceforth, `a` shall be a variable of type `float`!'.

Similarly, once you've stated

```
int b;
```

`b` will be an integer variable. If you subsequently do

```
b = 3.14;
```

trying to assign the value of 3.14 to the variable `b`, it simply 'won't fit'. What C will do is truncate it to an integer number (*i.e.* chop off everything after the decimal point – this is not the same thing as rounding!) so after this assignment, `b` will actually be 3. If you're lucky, the compiler will issue a warning, such as 'possible loss of data' (one reason to always compile with warnings turned on), but as far as C is concerned, this is not an error and it will happily continue. This can lead to rather unexpected results:

```
int a = 1;
int b = 3;
int c = a/b;
```

What do you think the value of `c` is after this piece of code? One third? You wish. Not even $0.333\cdots$ or anything like that. It's zero. This is something to be very aware of while programming, and we will devote some more attention to this in the next section.

Note that variable declarations must obviously appear in your program before the variable can be used. If you try this:

```
int main(void)
{
    a = 3.14;
    float a;
    return 0;
}
```

the compiler will complain with something like `'a' undeclared (first use in this function)`. You may, however, combine *declaration* with *initialization*:

```
int main(void)
{
    float a = 3.14;
    return 0;
}
```

Unfortunately, in C89 you may not declare new variables just wherever you like. Once 'other statements' have occurred in your program, you can no longer introduce new variables:

```
int main (void)
{
    float a = 3.142;
    printf("a = %f\n", a);

    float b = 2.718;
    printf("b = %f\n", b);

    return 0;
}
```

will cause the compiler to complain with an error message like `parse error before 'float'` followed by `'b' undeclared (first use in this function)`. The parse error basically means 'I wasn't expecting this word at this point in your program'. There is no good reason for this limitation of C (other than making compiler writers' life a little easier), and C99 removed this restriction (C++ does, too).

C knows more than just `floats` and `ints`. For example, there is also the type `char` (for *character*) which can hold a single byte (*i.e.* 8 bits). A *string* as we have seen before is nothing more than an array of `chars` – we'll get to arrays in a later chapter.

There is also the type `double` (short for 'double precision floating point') which offers better accuracy and a larger range. See §1.3.3 on page 23 for more information on range and accuracy. In fact, on most current 'ordinary' computer systems, there is no obvious advantage to using `float` instead of `double` (the CPU performs all floating point calculations in double precision anyway) so unless memory requirements dictate the use of `floats`, stick to `doubles` instead.

On older systems, an `int` used to be a 16-bit integer (and on some, it still is). C offers an integer type with increased range in the form of a `long int` (or simply `long`, for short) which usually is 32 bits. Similarly, to specify that you really want a 16-bit integer, you can specify `short int` (or `short`, for short). Since modern computers are not particularly bothered by working with 32 bit integers, there is rarely a reason to use `shorts`.

On some systems, there is also a floating point type available with extra precision, called `long double`. Beware that on systems not supporting this, it is actually the same as an 'ordinary' `double`.

Finally, you can prefix the integer types with an explicit `signed` or `unsigned`. Most types are implicitly `signed` (*i.e.*, they can contain positive and negative values), except for char which on some systems is unsigned by default. However, if you have a variable which can logically never be negative (for example because it represents a number of particles or something), or if you need the extra bit of range, you can use unsigned long.

C99 adds the _Complex and _Imaginary specifiers, so you can have, for instance, a double _Complex. The reason it is spelled in such a 'weird' way is because names starting with an underscore followed by a capital letter were always 'reserved', *i.e.*, you were not allowed to use them yourself (see §3.7). Using the word 'complex' for the new type could have broken existing programs. A complex type is stored as an array of two elements of the corresponding real type, the first being the real part and the second the imaginary part.

```
long double _Complex ldc = 42.0; /* imaginary part is zero */
```

If you #include <complex.h>complex.h, you get 'nicer looking' definitions for these, so that your code would look like

```
double complex dc = 42.0 + 37.3*I;
float imaginary fi = 2*I;
```

In this header file, there are also _Complex_I which is a complex constant with the value i, and _Imaginary_I which is an imaginary constant with the value i (or, if pure imaginary numbers are not available in this implementation, it is the same as _Complex_I). There's a helpful definition for I which is used in the examples above.

In complex.h there are also various mathematical functions – more on that later. We'll see an example of the use of complex numbers in §11.4.2.

Note that support for complex numbers is usually the last thing which is added to older compilers on their way to full C99 compliance, because it's a feature only used by a minority (unfortunately, we *are* a minority), and because it's a *lot* of work to make all the math libraries support complex arithmetic. Also, some systems already supporting _Complex do not yet support _Imaginary.

C99 also adds a _Bool type for booleans – more about that in §4.3 – and a long long (both signed and unsigned) which is at least 64 bits wide.

3.3 Expressions

In C, an *expression* is considered to be 'anything which evaluates to a value'. In general, everywhere a numerical value is expected, you can also type an expression. We've already seen that you can type

```
int a = 1;
int b = 3;
int c = a/b;
```

(Quick, what's the value of c after this code?) in which the 'a/b' part is an expression. The '1' and '3' are called *literals*.

An expression can be built up from literals, variables, and function calls (see below), although if an expression is built up entirely from literals, the compiler is smart enough to replace it with the result at compile time, so that when you do

```
int a = 1 + 4/2;
```

there would be nothing left to calculate at run time; the compiler replaces the entire expression with '3'. Note how the example was carefully chosen not to give any surprises with rounding fractions—we'll get to what happens otherwise below.

Expressions in C have a syntax common to most programming languages. The operators used are

math	C
$a+b$	`a + b`
$a-b$	`a - b`
$a \times b$	`a * b`
$a \div b$	`a / b`

Note that there is no operator for a^b. Some programming languages use `a**b` or `a^b` for that, but these mean something else in C. The asterisk (`*`) is used in most programming languages to prevent confusion with the letter 'x'.

The normal precedence rules of calculus apply, *i.e.* `1+2*3` is 7, not 9. You can add parentheses, so that `(1+2)*3` is 9. Note that spacing is of no significance and can even be misleading: `1+2 * 3` is still 7. Some programmers specifically write spaces around `+` and `-`, and not around `/` and `*` for this reason: `1 + 2*3`. This is also the convention used in this book.

Incidentally, C knows *many* more operators than just the four mentioned above. Most will be treated in this book, and a list of their precedence rules is given at the end of chapter 8.

Note that the mathematical convention that ab really means $a \times b$ does not work in C: `ab` is simply a variable named '`ab`', not somehow `a*b`.

Also note that you are free to use as many parentheses as you like, as long as they are matched (if you forget one, this will be flagged by the compiler as a syntax error). So `(((a + 1)*(((3.42) )) ))` is a valid expression (though a bit silly).

We will see more complex expressions once we have treated functions in more depth.

As promised, we return to fractions and rounding problems. If you type

```
double f = 2/3;
```

you might get a surprising result. Obviously, we are dealing with a floating point value here, so `f` will probably be something like $0.666\cdots$, right? Unfortunately not. It will still be zero. The reason is that the compiler *first* calculates the expression, which obviously involves two integers and hence will yield an integer result (which is always truncatedtruncation—in this case to zero), and *then* that value is assigned to the variable. This is arguably wrong—it would have been nice if an integer division yielded a floating point value—but alas, this is the way things are. It is even only in C99 that such truncation is defined to be 'towards zero'; in the previous C standards, it was 'implementation defined' whether truncation was 'towards zero' or 'towards $-\infty$'. Obviously this makes a difference when one of the values is negative.

C does make type changes when the expression itself contains multiple types. The rules followed by C when doing 'type promotion' are quite elaborate, by simply put, the 'most accurate' type used in an expression will 'win', and will determine the resulting type. For instance, adding an `int` to a `float` yields a `float`. Likewise, multiplying a `double` by an `int` will yield a `double`, etc.

So, to fix the example above, we would somehow have to make sure that the expression is evaluated to a `double`. There are two ways of doing so. One is to force the compiler into considering one of the values as a floating point value:

```
double f = 2.0/3;
```

or

```
double f = 2/3.0;
```

This works fine for literals, but not for variables. If you type

```
int a = 1;
int b = 3;
double c = a/b;
```

`c` will still be zero. The way to solve this is to use *type casting*:

```
int a = 1;
int b = 3;
double c = (double) a/b;
```

The `(double)` above means 'I know the following variable is really something else, but treat it as if it were a `double`'.

There is also a special notation popular among scientists, who often write numbers in the form 1.49×10^9. The 'times ten to the power of' is tersely spelled 'e' in C, so the value above could be entered as `1.49e9` in your program. Likewise, `43.21e-5` means 43.21×10^{-5}.

After this lengthy discussion about floating point and integer division, this is a good time to introduce a special operator, the *modulo operator*. It is special in that it only works on integer operands. The expression $p \bmod q$ is written `p % q` in C, and yields the remainder of the integer division p/q. I.e., `p % q` equals zero if q divides p.

If you try to use the `%` operator with non-integer operands, the compiler will say something like '`invalid operands to binary %`' and refuse to compile. For floating point operands, you can use the `fmod()` function: `fmod(7.3, 2.0)` yields 1.3. See the next sections for more on functions.

3.4 Functions

A function, in C parlance, is a piece of code with a name, taking zero (!) or more parameters, and optionally returning something. Just like your scientific view of functions (apart from the 'zero parameters' type, which we will get to later). You can 'call' a function from within your program; a statement like `x = sin(theta);` should look reasonably sane to you.

In your day-to-day use of mathematics, you don't specify exactly what kind of value comes out of a function; it is by context that you know whether a function returns a complex variable, an integer, or something else. In C, you specify exactly what type of parameters should go in, and exactly what type of parameters will come out.

In C, a function is defined like this:

```
out_type name( in_type_1 param_1, in_type_2 param_2, ...)
{
    'body': code to implement the function
}
```

A function can take as many parameters as you like, but can return only one. The return value is given simply by stating

```
return expression;
```

This expression can be a literal value, a variable, or 'anything which evaluates to a value', as was mentioned in §3.3. When the compiler finds this statement, it will leave the function and return to the line in your program immediately after the function call. An example might be in order.

Let's define a trivial function which calculates the square of the number passed to it. That function would look something like this:

```
double square(double x)
{
    return x*x;
}
```

What this piece of code does is define a function called `square` returning a `double` and taking a `double` parameter, which is called `x` and which is available in the 'body' (*i.e.* the code that makes up the implementation of the function) under that name. In this particular case, the function body is rather simple as it returns x × x.

Further on in your program, you can now use this function, as the following code shows (the interesting line has been set in boldface):

```
int main(void)
{
    double value, squaredvalue;
    printf("Please enter a value\n");
    scanf("%lf", &value);  /* "ell eff", not "one eff" */
    squaredvalue = square(value);
```

```
    printf("The square of %f is %f\n", value, squaredvalue);
    return 0;
}
```

For the moment, forget about the scanf line and take my word for it that it will allow the user of your program to type in a value which will be put in the double variable named value. What is interesting is that you can use square just like you would use a function like sin.

3.4.1 Format Strings

Now, about the strange %i in the printf statement. With that, you tell the printf function that you want to print an integer at that position in the string. It happens to be at the end of the string in our example, but it can appear anywhere, even multiple times:

```
int a = 42;
int b = 37;
printf("a = %i and b = %i\n", a, b);
printf("a + b = %i\n", a + b);
```

In the first printf, the first %i is replaced with the value of the first variable after the string (a), and the second with b (etcetera: you can have an unlimited amount of variables in one printf statement, should you wish). The second printf shows that you are even free to enter an expression instead of a single variable – just like with any function argument.

There is a whole slew of these magic percent sign combinations. The most often used ones are %f for a float or a double, %e to use the 'scientific notation' ('1.234e-4') and %E for the same but with a capital E; %s for a null-terminated string (a what? Don't worry, we will get to that in more detail later), and %c for a single char.

These too can be freely intermixed:

```
int a = 42;
double b = 3.14;
printf("a = %i and b = %f\n", a, b);
```

Note that many compilers don't notice if you type

```
int a = 42;
double b = 3.14;
printf("a = %i and b = %i\n", a, b);
```

i.e. say you're going to print an integer, but provide a double instead. Some compilers will issue a warning in this case, but this is considered 'advanced' for a compiler. We'll see why in the next section. If you ignore this warning and run your program anyway, you'll most likely get nonsense printed on the screen.

Also, C does not take notice of the 'accuracy' of a certain value. In science, when it is written that $x = 1.0$, one actually means $0.95 \leq x < 1.05$. But when you do

```
double a = 1.0;
double b = 1.0000;
printf("a = %f, b = %f\n", a, b);
```

both will be printed as `1.000000`. The number of digits used to print the value has nothing to do with 'significant digits' as we know them from science. You can change the amount of digits by putting a number between the percent sign and the type identifier: When the format is `%6.2f`, that means 'use at most 6 characters for this value, of which 2 after the decimal point'. The output is padded on the left side with spaces, so that

```
printf("%6.2f\n", 3.1415);
printf("%6.2f\n", 42.0);
printf("%6.2f\n", 123.456);
```

yields the output

```
  3.14
 42.00
123.46
```

which is a nice feature to neatly 'line up' values. Note that the output is correctly rounded if it doesn't fit in the number of digits. If you specify a leading zero, padding will be with zeros instead of spaces:

```
printf("%06.2f\n", 3.1415);
printf("%06.2f\n", 42.0);
printf("%06.2f\n", 123.456);
```

yields the output

```
003.14
042.00
123.46
```

You can also specify only the part after the decimal point, which means 'use however many characters you need, but only use this many digits after the decimal point':

```
printf("%.2f\n", 3.1415);
printf("%.2f\n", 42.0);
printf("%.2f\n", 123.456);
```

yields the output

```
3.14
42.00
123.46
```

If you specify nothing but `%f`, it is as if you had specified `%.6f`. Note that there is no provision to print out only the values after the decimal point that you have specified, without adding zeros. This is because C doesn't remember whether you have specified a value as being 1.0 or 1.0000 when giving the variable its value – both have the same internal representation.

If you need a percent sign itself, you can use `%%`:

```
printf("Current interest rate is %.2f%%\n", percentage);
```

Since format strings are often used and there are quite a few ways to format the data to be printed, here is a summary.

A format string contains two types of entries: 'ordinary' characters, which are output verbatim, and 'conversion specifiers', which determine how the corresponding argument of `printf` is converted and printed. Conversion specifiers begin with a `%` character and end with a conversion character. Between the `%` character and the conversion character, the following characters are allowed (among others):

- Flags (in any order) which change the specifier:
  `+`: always prepend the value with either a minus or a plus sign;
  space: if the first character is not a plus or minus, prepend a space;
  `0`: prepend zeros to fill the field width;

- A number signifying the field width in characters; the converted argument is printed in a field which has at least the given width (and more if necessary).

- A decimal point, separating the field width from the precision.

- A number specifying the precision. When the argument is a string, the precision is the number of characters to print; with a floating point number the meaning of 'precision' depends on the conversion character used. If it's `e`, `E`, or `f`, the precision determines the number of digits printed after the decimal point; if it's `g` or `G` it determines the number of significant digits. With integers, the precision determines the minimum number of digits to print (if necessary, prepended with zeros).

- A length *modifier*; `h` signifies that the corresponding argument is to be printed as a `short` or `unsigned short`; `l` for a `long` or `unsigned long`, or `L` for a `long double`.

The most-used conversion characters are listed in the table below:

Character	Type of argument; converted to:
d, i	int; decimal notation with sign.
x, X	int; hexadecimal notation.
c	int; one single character, after conversion to unsigned char.
s	char*; zero-terminated string.
f	double; decimal notation in the form [−]*m.dddddd*, in which the number of *d*s is determined by the precision (default is 6). When a precision of 0 is specified, the decimal point is not printed.
e, E	double; decimal notation in the form [−]$m.dddddde \pm xx$ or [−]$m.ddddddE \pm xx$.
g, G	double; use %e or %E if the exponent is less than −4 or greater or equal to the precision; else use %f. Zeros and/or a decimal point at the end of the number are not printed.
p	void*. Representation is implementation-defined (usually a hexadecimal notation of the integer value of the pointer).

The scanf function uses the same elements for its format string, with the exception that for a double argument, you need to specify %lf (ell eff) instead of just %f. You can memorize this by thinking of a double as a 'long float'.

3.5 Prototypes

In the case of your homemade square function, it is pretty obvious how the computer will know what to do when it is called. After all, the implementation of that function shows up in your source code just lines above. But for something like printf, how does the compiler know what to do?

This is where the mysterious #include <stdio.h> comes in. You might expect that it contains the implementation of printf (and a load of other functions) but that is not the case. Instead, it contains only the function *prototypes*. A function prototype is only a statement telling the compiler the name of a function, the number and types of its input parameters, and the type of its output value. They have the form of

outtype name(*intype*$_1$ p$_1$, *intype*$_2$ p$_2$, ...);

so in the case of a sine function which takes a real value and also returns a real value, the prototype would look something like

```
float sin(float radians);
```

When the compiler sees this prototype, it will 'know' from then on that the function sin takes exactly one parameter of type float, and will return a float. In that respect, a function prototype is a bit like the variable declaration we saw before. Strictly, it doesn't know *how* it should calculate the sine based on the input parameter, but that's of no concern. It will be the job of the linker to look for a function called sin, and make sure the corresponding code gets called.

Incidentally, the compiler doesn't *have* to have seen a prototype of a function before allowing you to call it. If it hasn't (most likely because you forgot to include the proper header file), you might get a warning such as

```
warning: implicit declaration of function 'sin'
```

but the compiler will assume you know what you're doing, insert the code to call the function with the parameters as you have specified them, and assume an `int` return type. As long as the linker can find the definition of the function, it will happily link your program into an executable.

This warning is one of the most dangerous ones to ignore, though. If you happen to have specified the parameters of the correct type, things will probably turn out right. If you specified the wrong ones (or not enough – in this case, just calling `sin()`) then the function will perform its calculations as if you *did* specify the parameters, but take its value from whatever happens to be on the stack at that time (most probably gibberish). We'll talk about this some more in chapter 10.

The compiler doesn't care about the *name* of the parameter(s), only about their type(s). The fact that it is called `radians` here serves as a reminder to the *programmer* that this particular version of `sin` expects its input in radians and not in degrees. No compiler will note it as an error when you pass this function an angle in degrees – it will happily assume that the value is in radians, and return the sine of the value as if it had been. To a human, seeing a parameter like '$k\pi$' will immediately flag it as being in radians and a value like 45 will hint at being in degrees, but to a compiler, a `float` is a `float` (is a `float`).

Now, it's time to re-visit the *header file*, as this is just a big collection of these prototypes (and perhaps some other stuff, but that is not important for now). So, if you `#include` such a header file, the compiler will suddenly 'know' lots of functions by name, what type of parameters they take, etc. For example, the file `math.h` contains the prototype for `sin` as well as a whole slew of other mathematical functions.

If you `#include <math.h>` in your program and try to use `sin("Hello")` somewhere, the compiler will complain with an error message such as

```
incompatible type for argument 1 of 'sin'
```

This is because it was expecting a `float`, and instead it found a string. The compiler knows of no way to convert a string to a floating point value, so it issues an error message.

This may seem a trivial thing, but this feature called *type-safety* is actually quite important. You probably wouldn't make the `sin("Hello")` mistake, but there are other, more subtle errors which would not be so obvious. The compiler knows of a few 'implicit conversions', so if you pass an integer parameter to the sine function, the compiler will silently convert it to a floating point value (since this can be done without problems). This is why you didn't get an error when you passed a `float` as the argument to `sin`, even when your particular version of `sin` was expecting a `double`: The compiler knows it can implicitly convert a `float` into a `double` without loss of information, so it will do so without even bothering to tell you.

However, the other way around is a different matter. If you have a function which takes an integer parameter and you feed it a floating point value, most compilers will issue a warning, since in general, a floating point value cannot be converted to an integer value without loss of information.

So far we have seen that you can use *functions* in your program. Even if the actual *implementation* of the function is unknown to you (or even to the compiler!), if you specify its *prototype*, it will be the job of the linker to find the code implementing that particular function and make sure it somehow ends up in your executable.

By the way, if after reading all this prototype stuff you are wondering how `printf` pulled it off to allow for a variable set of parameters (in which case you are quite astute): this is because the `printf` prototype is defined in a special way:

```
int printf(const char *format, ...);
```

Forgetting about the `const` (and about the `*`) for a second (both are material for later), the interesting part here is the '...'. This means 'a variable set of parameters', and tells the compiler that it should give up all parameter type checking at that point, leaving it up to the implementation. The big drawback here is that parameter checking has to be done at runtime (as opposed to compile time) which is the reason most compilers can't detect if you specify "`%i`" yet provide a `float`. It is also one of the things about C that programming language purists will never tire to point out as being a major flaw.[1] Luckily, most modern C compilers are able to 'look inside' the format string when compiling your code, and can issue warnings when the type of the argument doesn't match with the corresponding entry in the format string.

3.6 Into the Void

You have already encountered the mysterious `void` keyword a couple of times, and this is a good time to explain what it is about.

Suppose you have a function that returns some value but does not take any input parameters. The mathematicians amongst our readers would probably raise their eyebrows at such a strange 'function', but there are plenty of examples of where such a function would be useful. One would be a 'random generator' which would return a different number each time it is called—useful for simulations (and games). Such a function does not take any parameters, and the prototype would be something like

```
int random(void);
```

where `void` means 'this function does not take any parameters'. Such a function can be used in your program like so:

```
int a = random();
```

You see that the function call still uses the parentheses—this is how you can easily tell apart a *variable* from a *function* (in day-to-day mathematics, you can write something like '$y = \sin x$' which means the same as '$y = \sin(x)$'—there, typography also helps you distinguish function names from variables. This is not the case in C).

[1] If you are ever bullied about this by a Pascal fan, ask that person to provide a `writeln()` implementation in Pascal. At least you can write a C compiler in C.

By the way, you would expect that you should have been able to declare your zero-parameter functions like so:

```
int random();
```

(after all, this is how you call it too), but for historical reasons this does not work—you will have to explicitly tell it that it's `void`. (In C++ it is valid to omit the `void`.)

There is also the possibility of a function that has no return value. This is mathematically even stranger, but it is what other programming languages call a *subroutine* or a *procedure*. Basically, you can view a procedure as a subsection of the program which you have given a certain name, so that you can refer to it elsewhere by that name. The steps in our 'recipes', earlier on, were such procedures.

Again, you may expect that you could define such a function like so:

```
do_something(void)
{
    do something here
}
```

but alas, if you leave away the return type of a function, it defaults to `int` instead of `void` (in C89 that is – in C99 it has been made illegal). Instead, you would have to say

```
void do_something(void)
{
    do something here
}
```

In this particular case, the function `do_something()` also does not take any input parameters, but you can imagine procedures that take an input parameter yet do not return anything.

Note that there need not be a `return` statement at the end of your procedure: If the computer reaches the end of it, it will understand that it needs to return to where it was when the procedure was called, and pick up from there.

3.7 Names

In C, you cannot name your variables and functions just anything you like. A name (also called *identifier*) must start with a letter or an underscore (_). But after that, any letter, underscore, or digit is allowed (provided you don't pick one of the 'reserved words'). So, a few examples of correct names are

- `a`
- `foo`
- `SomeNameWithCapitalLetters`
- `a_long_name_with_underscores`

- `foo123`
- `_a_name_starting_with_an_underscore`

The latter deserves a special caveat: You can start names with an underscore, but you shouldn't. Names starting with an underscore, followed by a capital letter (and names starting with two underscores) are reserved for the C compiler's internal stuff and for the operating system.

Note that in C, names are case-sensitive. That is to say, `Foo` is different from `foo` and `FOO`. Some examples of incorrect names:

- `123abc` (does not start with a letter)
- `Some Name` (contains a space)
- `Argh!` (contains an exclamation point)
- `me@mycompany.com` (at-signs and periods are not allowed either)

A word of advice: Try to give your variables and functions meaningful names. Especially when your programs get bigger and more complex, it gets harder to keep track of what names mean. Also, when you look at a program you wrote a few months ago (or even worse, which somebody else wrote), it can be tough to figure out what `calculate()` and `find()` did again (you would have less of a hard time when you had called them `calculate_energy()` and `find_zero()` for example). On the other hand, typing long names can be tedious, and it is debatable whether '`number_of_items`' is really more descriptive than '`n`'. Also, '`Ekin = m*v*v/2`' to any scientist is just as easy (if not easier) to read as '`kinetic_energy = mass*velocity*velocity/2`'. You will find a balance naturally.

Incidentally, C99 allows for non-ASCII characters in identifiers (see §5.3.1), such as greek, Thai, or Cyrillic characters.

The 37 'reserved words' in the C language are:

```
auto      enum      restrict  unsigned
break     extern    return    void
case      float     short     volatile
char      for       signed    while
const     goto      sizeof    _Bool
continue  if        static    _Complex
default   inline    struct    _Imaginary
do        int       switch
double    long      typedef
else      register  union
```

Most (but not all) of these will be covered in this book.

3.8 Comments

Giving your variables sensible names is helpful to understand what a program is supposed to do. But for complex programs, this is not enough. Therefore, you should add *comments* to your code. In C, everything standing between /* and */ is considered a comment and is ignored by the compiler. Comments may span multiple lines:

```
/* Calculate the next cardinal
   Note: should really use long ints */
int n_plus_one(int n)
{
    return n + 1;
}
```

Most compilers also allow 'C++-style' comments, which start with a double slash (//) and run until the end of the line. This is an 'official' addition in C99.

Adding sensible comments to your code is sometimes more of an art than writing the actual code itself. Overzealous commenting leads to stating the obvious:

```
a = a + 1; // Increase the value of a by one
```

and on the other end of the spectrum are comments which add nothing to the understanding of the program (often added as an afterthought by programmers because 'commenting is mandatory'):

```
/* The following function calculates the value and returns it */
```

Clearly, this is of not much use.

As a guideline, you should add a block of comment at the top of your program stating what it is supposed to do and immediately before every (non-trivial) function. Think 'what would I be interested to know about this function if someone else would have written it?' Things like valid input values, what will happen if you feed it an *invalid* input value, etc. Also, terse comments in the actual code, especially when doing something complex, can be a good idea.

3.9 More about Math

Now that we know how to write mathematical expressions in C and how to call functions, we can take some time to look at more complex mathematical expressions and a few often-used functions that are available. In the `math.h` header file (which comes standard with each C installation) there are a whole bunch of useful mathematical function prototypes. For example, you may remember that C has no operator to express x^y, but you can get the same effect with the function `pow(x, y)` (both parameters are of type `double`, like most parameters in `math.h` functions). Also, to calculate $\sqrt{x}$, you can use `sqrt(x)`.

There are also the standard functions `sin`, `cos`, and `tan`, as well as `sinh`, `cosh`, and `tanh` for the hyperbolic versions; there is `asin(x)` for $\arcsin x$, `acos(x)` for $\arccos x$, and `atan(x)` for $\arctan x$. The latter also has a special version `atan2(x, y)` for $\arctan \frac{x}{y}$ (useful when $y \approx 0$).

Also, there is `exp(x)` for e^x, `log(x)` for $\ln x$ and `log10(x)` for $\log_{10} x$.

Each of these functions can be used in compound expressions. A few examples:

$\cos^2 \theta$	`pow(cos(theta), 2)`
$\sqrt{x^2 + y^2}$	`sqrt(pow(x, 2) + pow(y, 2))` or simply `sqrt(x*x + y*y)`
$\arctan {}^1\!/\!\sqrt{x}$	`atan2(1, sqrt(x))`

In C99, the header file `complex.h` includes various functions for use with complex numbers, including

- trigonometric functions (`csin`, `ccos`, `ctan`, `casin`, `cacos`, `catan`)
- hyperbolic functions (`csinh`, `ccosh`, `ctanh`, `casinh`, `cacosh`, `catanh`)
- exponential, power, and logarithmic functions (`cexp`, `cpow`, `csqrt`, `clog`)
- absolute value and complex manipulation (`cabs`, `creal`, `cimag`, `carg`, `conj`, `cproj`)

Each of the functions above takes a `complex double` argument, and there are versions taking a `complex float` (with the function names having a `f` appended, such as `csinf`), and versions taking a `complex long double`, with an `l` suffix (such as `csinl`).

Note that there is an important limitation: C will do *calculus*, but not *algebra*. There is no provision for things like integration and differentiation. Given a function $f(x)$ you cannot find $\frac{df}{dx}$ by simply typing `df/dx`, nor can you express $\int_0^x f(x')dx'$ in C in any way. There are several ways of calculating the *numerical value* of such expressions (and we'll cover a few), but you cannot (simply)[2] evaluate to symbolic expressions.

Another important thing to note is what happens when your calculations yield results which cannot be represented by variables in C. For example, when you have an integer division by zero

```
int q = 0;
int p = 1/q;
```

the processor, when executing this code, will generate a so-called *exception*, since it cannot provide a sensible answer. The operating system will catch this exception and terminate the offending program. To an outsider, it will simply look as if your program has *crashed.*

For floating point variables, the situation is slightly different. The binary representation of floating point numbers (see §1.3) has special cases so that a floating point number can have the values ∞, $-\infty$, or 'NaN' (not-a-number). Any value which is 'too large' to be represented by a floating point number, will be 'rounded to ∞', so to speak. 'NaN' is reserved for 'undefined' values, such as $0 \times \infty$. Exercise 3.7 investigates these special values.

[2]Of course, there *are* programs which *can* do symbolic algebra, and you can write one in C. But this is several orders of magnitude more difficult. Nobel prize difficult. Recently, Veltman shared the Nobel prize for physics with 't Hooft, in part thanks to writing the first program ('Schoonschip') which could deal with symbolic algebra to help solve some hideous equations.

3.10 Return Values and Error Codes

Now that we have seen that functions (in C parlance) can take any number of parameters (including zero) and can return either one value or none at all, it is time to point out that `void` functions (*i.e.* without a return value) are relatively rare in C. This is because in the case of a routine which does not 'calculate' anything but only 'does' something, the return value is often used as a success indicator or error code. For this, usually a simple `int` is used.

For example, consider a `print_all_results()` routine that sends all the results of some calculation to a printer. You would call that routine at the end of your program, for example. But during printing, there are things that can go wrong: The printer may be out of paper, for example. So instead of declaring that function to be `void` and hoping for the best, you could make it return an `int` which encodes the status of the printer.

```
int print_all_results(void)
{
    if (the printer is out of paper)
        return 1;

    print all results
    return 0;
}
```

This is a rather smart thing to do. This way, you can make your programs 'self-checking' while they are running, and by defining certain error codes for certain errors and sticking to them, you can figure out what went wrong during a program at all times. It takes some getting used to, but if you stop and think for a moment what can possibly go wrong during execution of a particular routine you write, and define error codes to match those situations, you will be able to trace problems at run time much easier.

It is convention in the world of C programming that a return value of zero means 'success'. Also, it can get rather confusing to memorize what all the error codes mean – in this case, 'one' meant 'out of paper', but every time you use that `print_all_results()` function, you have to remember that. There is a handy feature of the C preprocessor which lets you assign human-readable names to numbers (or anything else, for that matter) by using the `#define` directive. So, if you put somewhere at the top of your program

```
#define SUCCESS        0
#define OUT_OF_PAPER   1
#define OUT_OF_INK     2
```

then you can refer to these error codes with these names. By convention, these definitions are written in all capitals so they stand out in your code. A good place for such definitions would be in the header file which also contains the prototype for your functions. When you have included this header file, you can do things like the following. Forget about the if-statement for a second (it does what you think it would) and about the strange '==' in there. We will get to that in the next chapter.

```
int error = print_all_results();
if (error == OUT_OF_PAPER)
    printf ("Please put some paper in the printer\n");
if (error == OUT_OF_INK)
    printf ("Please put a new ink cartridge in the printer\n");
```

This increases the readability (and understandability) of your code considerably. Note, however, that such a #define is just another name for whatever you decided to give that name, and the compiler will never even see that name (it is replaced with the actual definition by the preprocessor). Therefore, nobody will prohibit you from writing things like

```
int a = OUT_OF_PAPER*17;
```

which is clearly doesn't make much sense. Use your good judgment; using proper #defines can make your life much easier, but will not prevent you from doing silly things.

3.11 Macros

The constants defined with the #define preprocessor directive, which we saw in the previous section, are called *macros*. It is important to realize that the compiler never sees them; they are replaced by their definition by the *preprocessor* (which, as its name suggests, runs *before* the compiler)[3].

These macros need not be simple constants; they can actually look a bit like functions. For example, you could define

```
#define SQUARE(x) (x)*(x)
```

and use this macro just like it were a function. One reason programmers sometimes do this is because it's more efficient: a 'real' function call incurs some performance overhead, and the definition of a macro is inserted straight into the code. However, there are some caveats. You may wonder why the definition above used these 'unnecessary brackets', but consider the following:

```
#include <stdio.h>
#define BADSQUARE(x) x*x

int main(void)
{
    printf("%d\n", BADSQUARE(2 + 5));
```

[3]This is actually an argument *against* using macros. When you use a *debugger* (a program which lets you step through your executable while displaying the corresponding location in the source code, which makes it easier to find where something goes wrong) then everything which was replaced by the preprocessor is 'invisible' to the debugger. Debuggers are briefly explained in chapter 10.

```
    return 0;
}
```

You may expect to see 49 printed out (7^2) but instead, you will see 17! What's going on? The reason is that macro expansion is taken *very* literally, so if `x` is '`2 + 5`' (note the lack of brackets), then `x*x` will be expanded to '`2 + 5*2 + 5`' which is 17.

3.12 The Main Thing

Finally, there is one special thing that deserves your attention. For C, the `main()` function is a function like any other, with only two 'special treats': Firstly, you don't need to make a prototype for it, since every C program must contain a `main()`; and second, it is automatically called when you run the program. What happens in fact is that the operating system calls your `main()` function[4].

Now, we have been using `int main(void)` all the time. This `int` return value is passed to the operating system at program termination, and can be used to designate 'success' or 'failure' of your program. This can be useful if you take a step back and view your entire program as 'just another routine', which could possibly be accessed from other programs, or from a script or batch file.

Again, the convention is that 'zero' means 'success', which is why we ended all our examples above with '`return 0;`'. Almost all programs on UNIX systems stick to this convention; unfortunately this is less the case on Windows.

You may be wondering whether you can make a `main()` which *does* take parameters, and where those parameters would come from. The answer is yes, you can, and the parameters come from the operating system. We will get to that later. First, we will have a chapter about `ifs` and `elses` (no buts).

3.13 Synopsis

A *variable*, just like in day-to-day mathematics, is simply a *named entity* which can change during the course of your program. Variables have a specific *type* which determines what *kind* of values they can hold: integers, floating point values, etc.

A *function* is a separate piece of code which takes zero or more parameters and returns either one result, or nothing at all. The types of the parameters and the return type of the function are defined with a *function prototype*. To designate 'nothing', C uses the 'pseudo-type' `void`.

An *expression* is built up of values, variables, function calls, and/or operators. The normal operator precedence rules from calculus are in effect, and you can use parentheses in expressions to override the normal precedence rules.

[4]Actually, this is done by the 'glue' which is linked into your program.

Comments are pieces of text in the source code between `/*` and `*/` markers or (as of C99) everything on a line after `//`. Comments are ignored by the compiler, and are only there for the (human) reader's information.

When using functions as *subroutines*, *i.e.* pieces of code which are used multiple times and therefore broken out into their own section and which usually do not have to return a value, it can be useful to use the return value as an *error code* (usually zero in case of success) to notify the caller of the subroutine about whether something went wrong.

The *preprocessor* can be used to define *macros*, which can be simple constants but also function-like definitions taking parameters.

Lastly, doing *numerical calculations* is easy in a C program, but doing *symbolic algebra* is a whole other thing, for which C is rather unsuited.

3.14 Other Languages

Most other languages have the concepts of *functions* and *variables*. Only some very old BASIC dialects lack the concept of a function; they only have 'subroutines' (*i.e.* 'functions' taking and returning `void`). Some languages (notably Pascal) make an explicit distinction between functions and subroutines.

Type safety is not quite taken equally serious by all programming languages. Some, like Pascal, are quite adamant in enforcing it; others, such as older FORTRAN dialects, will not catch it if you pass an integer value to a function expecting a floating point value (called a '`REAL`' in FORTRAN). FORTRAN also has a feature where the type of a variable is automatically inferred from its first letter: You can set it up in such a way that all variables starting with `I` – `N` are integers, and the rest are `REAL`s.

BASIC, in principle, knows only two types of variables: 'numbers' (`float`s, in effect) and 'strings' (the equivalent of `char` arrays, which we'll get to in a later section). It offers basic type safety in that you can't add a 'number' to a 'string'.

In many scripting languages, most notably javascript, you actually *can* add a number to a string. In that case, the textual representation of the number will be concatenated to the string. This leads to interesting features:

```
"Adding 2 and 2 yields " + 2 + 2
2 + 2 + " is the result of 2 plus 2"
```

The former line will result in `"Adding 2 and 2 yields 22"` but the second one will result in `"4 is the result of 2 plus 2"`.

As for structure, most interpreted languages simply start executing your program from the top down – they do not look for a special `main()` function. In Pascal, you first define all your functions and subroutines, and at the end of your program make a `Program` section which contains the main code. This top-down structure is a heritage from the fact that Pascal compilers can compile a Pascal program in one single pass (which has the benefit of very fast compiles).

Some languages have different restrictions on names (for instance, FORTRAN77 requires that variable names be unique in their first six characters), while Pascal is not case-sensitive (*i.e.* `Foo` and `foo` designate the same variable).

As for error handling, some languages have a mechanism like 'on error go to ...', allowing you to specify a general error handling routine; other languages (such C++, C#, and Java) have the concept of *exceptions* which you can 'catch' in your program.

Some other languages have a preprocessor too (most notably C++, which actually uses the same preprocessor as C does) but generally, using it for anything other than `#include`ing other files is frowned upon.

3.15 Questions and Exercises

3.1 Since `void` is a valid 'return type' for a function, do you think you can also put

```
void a;
```

as a variable declaration in a program? Try it, and see what the compiler thinks of that idea.

3.2 If you do

```
double a = 1;
int b = 2;
double c = a/b;
```

what do you think the value of `c` is?

3.3 Write a function `cube()` which takes a `double` parameter and returns the cube of this parameter.

3.4 Write a function to return the hypothenusa $\sqrt{x^2 + y^2}$ of its `double` parameters `x` and `y`. (Many C distributions include such a `hypot()` function already, as it is standardized in C99.)

3.5 C offers $\ln x$ and $\log_{10} x$, but not a generic $\log_n x$. Write one.

3.6 Write a pair of functions, one to convert a temperature in degrees Centigrade to degrees Fahrenheit and one for the other way around. (If you call these `c2f()` and `f2c()` rather than `centigrade_to_fahrenheit()`, you have captured the C spirit quite well.)

3.7 Write a program to examine the 'special case' values for floating point variables mentioned in §3.9. If you use `printf` to print out these values, they will usually be represented as '`inf`' and '`nan`'. You can 'generate' ∞ by multiplying a value by a large number (say, `1e10`) several times, and printing it out each time. Find out what C thinks is the value of $\frac{\infty}{\infty}$, and especially of $\frac{1}{0}$. Is the latter mathematically correct?

Questions and Exercises

Chapter 4
Program Flow

Upon those who step into the same rivers
different and ever different waters flow down.

Heraclitus of Ephesus

4.1 Decisions

Suppose you are writing a function to calculate the absolute value of its parameter: $x \rightarrow |x|$. The code for such a function would simply return x if x is positive or zero, and $-x$ if x is negative. But how can we do that? How can we execute one piece of code for negative values, and another piece for positive values?

For this purpose, C provides an `if`-statement and an `else` statement. In our case, our function would look something like this:

```
int abs(int x)
{
    if (x < 0)
        return -x;
    else
        return x;
}
```

In general, the syntax looks like this:

```
if (condition)
    statement or block executed if condition is met
else
    statement or block executed otherwise
```

The `else`-part is optional, meaning it can be omitted if there isn't anything to be done in case the condition isn't met.

Note that there is no semicolon after the `if` and `else` statements! See §4.5 for more information.

You might have wondered about the 'statement or block' part above. A 'block' is a group of statements enclosed by curly braces. If this sounds familiar to the definition of the 'function body' in the previous chapter, that is because a function body is such a block. This is useful when you have more than one thing to do when a certain condition is met. Suppose we not only want to return the absolute value of the parameter for our `abs()` function, but also print out whether the original value was positive or negative. We will have to group stuff in blocks then:

```
int abs(int x)
{
    if (x < 0)
    {
        printf("Value was negative\n");
        return -x;
    }
    else
    {
        printf("Value was positive or zero\n");
        return x;
    }
}
```

By the way, note how we used indentation to make the code more readable. This is just 'good practice', and C in no way enforces it. The program above might just as well have been formatted like so:

```
int abs(int x)
{
      if (x< 0) {
        printf("Value was negative\n");
    return -x;} else
        { printf  ( "Value was positive or zero\n") ;
    return x;}
}
```

or even uglier. To C, it doesn't matter. To the human reader, it does. Don't do it. Keep your code tidy and readable. It is a matter of taste whether the first opening curly brace should be on a line of its own or be appended at the end of the previous line, yielding slightly more 'compact' code:

```
int abs(int x) {
    if (x < 0) {
        printf("Value was negative\n");
        return -x;
    } else {
        printf("Value was positive or zero\n");
        return x;
    }
}
```

For really short and simple `if`-statements, sometimes you see people write the statements on one line:

```
int abs(int x)
{
    if (x < 0) return -x;
    else return x;
}
```

Since this is all really a matter of (good) taste, just make one choice and stick to it. Therefore, a section titled 'Decisions' was a good place for these preponderances.[1]

So far, we have only seen single `if-else` statements, but you can also 'chain them together'. Suppose you want to make a function that tells you whether a given value lies within a certain interval, or to the left of it, or to the right of it. That would look something like this:

```
void check_interval(double lower, double upper, double value)
{
    if (value < lower)
        printf("Value lies to the left\n");
    else if (value > upper)
        printf("Value lies to the right\n");
    else
        printf("Value lies within interval\n");
}
```

4.2 Switches

Suppose you have a function which has several ways to compute its result, based on a parameter 'switch' which is only known when the program is run (for example, because it's a parameter the user has to supply when running the program). Consider the following interpolation example.

```
#define INTERP_NEAREST  1
#define INTERP_LINEAR   2
#define INTERP_CUBIC    3

double interpolate(double x, int method)
{
    double value;
    if (method == INTERP_NEAREST)
    {
        /* calculate value at nearest neighbor of x */
        value = nearest_neighbor(x);
```

[1]There are international contests for the most 'obfuscated' C program, if that's your thing – see `http://www.ioccc.org`.

```
    }
    else if (method == INTERP_LINEAR)
    {
        /* calculate value using linear interpolation */
        value = linear_interpolation(x);
    }
    else if (method == INTERP_CUBIC)
    {
        /* calculate value using cubic interpolation */
        value = cubic_interpolation(x);
    }
    else
    {
        /* error: unknown interpolation type */
        printf("Huh?\n");
    }
    return value;
}
```

For 'switch' types like these, C has a special construct aptly named `switch`. The above example would look like this:

```
#define INTERP_NEAREST  1
#define INTERP_LINEAR   2
#define INTERP_CUBIC    3

double interpolate(double x, int method)
{
    double value;
    switch (method)
    {
        case INTERP_NEAREST:
            value = nearest_neighbor(x);
            break;
        case INTERP_LINEAR:
            value = linear_interpolation(x);
            break;
        case INTERP_CUBIC:
            value = cubic_interpolation(x);
            break;
        default:
            printf("Huh?\n");
    }
    return value;
}
```

You see four new C keywords here. The `switch` takes an integer expression and is followed by several `case`s. Mind the colon at the end of the `case` line. Note that you can only switch on integer expressions. Each case in the above example is ended by a break statement which will terminate the switch. If you don't, the computer will continue with the next case. This is handy if there are several cases with common code:

```
#define INTERP_NONE    0

switch (method)
{
    case INTERP_NONE:
        /* No interpolation - we'll take nearest neighbor
           interpolation instead - so no break here... */
    case INTERP_NEAREST:
        value = nearest_neighbor(x);
        break;
    case INTERP_LINEAR:
        value = linear_interpolation(x);
        break;
    case INTERP_CUBIC:
        value = cubic_interpolation(x);
        break;
    default:
        printf("Huh?\n");
}
```

This turns out to be almost never the case, so in retrospect it would have been better if the next case statement would automatically break, *unless* it ended with a statement such as continue (which is already a C keyword). As it is now, you are bound to forget a break at least once in your life. Suppose the INTERP_LINEAR case above would not end in break; the computer would then continue with the INTERP_CUBIC case (which would overwrite value with a new calculation). Such errors are very hard to spot.

The last new keyword you saw was default. If none of the other cases match, the code in the default case is executed.

Finally, a word of caution. It is easy to 'abuse' the switch construct for a bad style of coding:

```
void doit(int what)
{
    switch (what)
    {
        case 1:
            /* do something */
            break;
        case 37:
            /* do something completely different */
            break;
        case 42:
```

```
                /* send an email to all your friends */
                break;
            case 123:
                /* format your hard drive */
                break;
            default:
                /* blink the screen */
        }
    }
```

In other words, it facilitates bundling unrelated code. This is a bad thing as it leads to code which is hard to understand.

There is one last flow control construct in C, the `goto`. But the mere mention of that word will make computer scientists start foaming at the mouth because it is considered a recipe for unstructured programs (the aforementioned 'spaghetti code').

4.3 Boolean Expressions

In the previous section, we said that you can use the `if`-statement to check for a certain condition, such as '`x > 0`'. This is half the truth. The 'condition' between the parentheses is really a *boolean expression*. A boolean expression is an expression which evaluates to either 'true' or 'false'. Nothing more, nothing less. In C, there is no 'maybe' or 'sometimes'.

That being said, C does not *really* support boolean algebra. A boolean variable should strictly be a type of its own, separate from `float` and `int` for example, which could then hold only the values `true` and Tndexfalse. In C++, for example, this is the case: There is a separate type `bool` with precisely these features. In C99, the `_Bool` type was added (C++ predates C99); more about `_Bool` later.

In C, an expression is considered 'false' when it yields (exactly) zero, and 'true' otherwise. Therefore, you often see C programmers do

```
#define BOOL int
#define FALSE 0
#define TRUE 1
```

although there are some caveats (for example, an expression yielding 2 is also considered 'true', but comparing it to `TRUE` wouldn't work as expected—we'll get to this in more detail below). As mentioned in the previous chapter, it is customary to give your own definitions capital letters so you can easily spot them in your code.

For boolean algebra, the operators 'and', 'or', and 'not' are important. They mean what you'd think they would: 'p and q' is true if both p and q are true; 'p or q' is true if p is true or q is true (or both), and 'not p' is true if p is false (and vice versa). In C, those operators are spelled `&&`, `||`, and `!`, respectively. The examples above would look like '`p && q`', '`p || q`', and '`!p`', respectively. (That is not a typo: the `&` and `|` really need to be typed twice.)

With these, you can construct complex condition expressions. For example, consider a function that checks whether a value is within a certain range (boundaries exclusive, but that's a detail):

```
BOOL in_range(double lower, double upper, double value)
{
    if (lower < value && upper > value)
        return TRUE;
    else
        return FALSE;
}
```

(using the definitions mentioned above.) After this, you can do something like

```
if (in_range(0, 1000, x))
{
    The value x is in the range
}
```

When using compound expressions like this (using the `&&` or `||` operators), it is important to know about a feature of C called 'lazy evaluation' or 'short-circuiting'. If you use the expression `p && q` and `p` is already false, then `q` is never evaluated (since the total result can never be true anymore). Likewise, if you use `a || b` and `a` is already true, the resulting expression is true and `b` is never evaluated. C is *defined* to do lazy, left-to-right evaluation. This is important because it allows you to write expressions like

```
if (x > 0 && tan(x) != 42) ...
```

in languages which do not promise left-to-right evaluation and short-circuiting (such as Pascal), the `tan(x)` part might be evaluated even if `x` = 0, which would trigger an overflow exception.

Note that the `in_range()` function can be written more compact, since the expression in the `if`-statement already *is* a boolean expression:

```
BOOL in_range(double lower, double upper, double value)
{
    return lower < value && upper > value;
}
```

You can also test for equality, and herein lurks *the* most frequently made mistake among new C programmers. Comparing two values must be done with the '==' operator—that's *two* ='s:

```
if (a == b)
{
    The values of a and b are equal
}
```

This would not have been such a problem if forgetting to put the second '=' would simply be a syntax error, but it's not. The following is valid C:

```
if (a = b)
{
    But what does it mean?
}
```

This means 'assign the value of b to a, and if the resulting value is non-zero, then...' This is almost never what you intended, but the compiler can only assume that you know what you're doing. In most cases, you can configure the compiler in such a way that it issues a warning in this case, but it is definitely something you *will* get bitten by. For this reason, some people insist on writing `if (0 == a)` instead of `if (a == 0)` because in that case, forgetting one of the ='s *will* constitute a syntax error ('assign the value of a to the number zero'?). Unfortunately, this only works when comparing to numerical values, not in the `a = b` case above.

It is easier to remember if you consider that 'and' and 'or' are spelled with double symbols ('&&' and '||') as well. Incidentally, using a single & or | is also valid C and means something else, which will be the subject of chapter 8. This all may look like C was deliberately designed to make it more error-prone, but this strange syntax is due to historical reasons. Again, you most probably *will* make this mistake at least once in your programming career. I apologize in advance.

Let us expound a little bit on the caveat mentioned earlier, that boolean expressions *yield* either 0 (false) or 1 (true), but in C, every non-zero value is considered 'true'. Therefore, it is safe to compare a boolean value to FALSE as defined above, but not to TRUE, since every non-zero value is 'true' but not every non-zero value is equal to 'TRUE'. Confusing? It can get even more fun when comparing booleans to each other.

Suppose you want to have a boolean function taking two booleans, and returning 'true' if both parameters are 'true' or if both are 'false', and 'false' otherwise. In other words, you want to implement the following truth table:

a	b	$f(a,b)$
true	true	true
true	false	false
false	true	false
false	false	true

Your first try at such a function might be something like this (remember, '!a' means 'not (a is true)', *i.e.* 'a is false'):

```
BOOL f(BOOL a, BOOL b)
{
    if (a && b)   return TRUE;
    if (a && !b)  return FALSE;
    if (!a && b)  return FALSE;
    if (!a && !b) return TRUE;
}
```

This is perfectly fine, but can be written in a more compact way:

```
BOOL f(BOOL a, BOOL b)
{
    return (a && b) || (!a && !b);
}
```

This reads as 'return whether either a and b are both true, *or* a and b are both *not* true'. The extra pairs of parentheses around 'a && b' and '!a && !b' are to increase readability. Incidentally, you are free to add as many pairs of parentheses as you like, as long as they are matched. If not, the compiler will detect this as a syntax error.

Now, you may be tempted to write this even more compact. After all, the function should return 'true' when a and b are *the same*, so why not explicitly say so?

```
BOOL f(BOOL a, BOOL b)
{
    return a == b;
}
```

It can't get more compact than that! Unfortunately, it may also not quite do what you'd expect—that is to say, it is not entirely equivalent with the previous implementations. If you call this function with parameters, say, 1 and 37, both of which should be considered 'true', yet $1 \neq 37$ and thus this version of the function would return 'false'.

By now, you probably have a healthy distrust of self-defined BOOL types, and that is a good thing. It's certainly not for nothing that _Bool was added in C99.

The reason this is spelled with the underscore and the capital letter is the same as for _Complex and _Imaginary: Names like these were always 'reserved' (see 3.7) so adding them to the language later would not 'break' any existing programs. A _Bool can hold exactly two values: 0 and 1 (for 'false' and 'true', respectively). Assigning anything other than 0 to a _Bool automatically converts it to 1, so it *is* safe to compare _Bools. If you #include <stdbool.h> (which was also added in C99), you will get the nicer-looking bool type (which is the same as _Bool), and true and false which are defined as 1 and 0, respectively. This is the same syntax as in C++.

Apart from the comparison operators <, >, and ==, there are also <= (for $\leq$), >= (for $\geq$), and != (for $\neq$). There are no comparison operators for $\ll$ ('much smaller than') or $\gg$ ('much bigger than'). After all, how much 'much' is, is completely arbitrary. The operators << and >> *do* exist, but they mean something different (see chapter 8). Also, there is no operator for $\approx$, since 'approximately the same' is not well-defined either. Sometimes you would wish there was, because of the following.

A big caveat involves comparing floating point values. At times, the result of a calculation involving floating point variables can have an unexpected result:

```
#include <stdio.h>
int main()
{
    float a, b;
    a = 1.0/3;
    b = 3.0;
```

```
        if (a*b == 1.0)
            printf("One is still one...\n");
        else
            printf("What's this?!\n");

        printf("a*b - 1 = %e\n", a*b - 1);
        return 0;
    }
```

Try out this piece of code, and see what it prints out. Since $\frac{1}{3} \times 3 = 1$, you would expect to see 'One is still one...'. Surprisingly, you don't. The reason is that floats have finite precision, so a is not *really* $\frac{1}{3}$, but some approximation. Therefore, a*3 is not *exactly* 1, and == only cares about *exact* matches. The last printf prints out the difference between this approximation and the 'real' 1 (using the %e notation here, because the difference is very small and would otherwise be printed as '0.000000').

So, comparing floating point values for equality is a dangerous thing. Instead, one should see whether two values are 'within ε'. You could do that like so:

```
if (fabs(a - b) < epsilon)
{
    /* Your code here */
}
```

of course, epsilon should then be another floating point variable, containing a value like 1e-6 or thereabouts, depending on how close you want a and b to be.

4.4 Loops

Computers are particularly well-suited for repetitive tasks. They don't get tired and they don't get bored. But how can we get them to do a given thing repeatedly?

Suppose your teacher wants you to add the numbers 1 through 100. Of course, every kid since Gauss knows that the answer is 5050 (or more general, the sum of 1 through N is $\frac{1}{2}N(N+1)$). Suppose you want to write a program that *really* adds these numbers anyway. You could start typing like this:

```
int sum = 0;
sum = sum + 1;
sum = sum + 2;
sum = sum + 3;
/* Etcetera */
```

This can't be a good idea. Even worse, you need to write a new program for each N. Luckily, C offers several constructs to execute a piece of code several times, in a so-called 'loop'. We will start with the simplest; you can probably read this straight away:

```
int sum = 0;
```

```
int n = 1;
while (n <= 100)
{
    sum = sum + n;
    n = n + 1;
}
printf("The sum of 1 through 100 is %i\n", sum);
```

This code works exactly as it reads: It will first set the variable `sum` to zero and `n` to one, then it will execute the piece of code between { and } repeatedly as long as the condition `n <= 100` holds. Since we increase the value of `n` in each iteration, it will become 101 after 100 iterations, at which point `n <= 100` is no longer true, and your program will continue with the first statement after the loop.

Since statements like '`sum = sum + n`' are so common in C, this is a good time to introduce a nice C syntax shortcut: You can also write '`sum += n`' for this statement. So, the loop above could also be written as

```
while (n <= 100)
{
    sum += n;
    n += 1;
}
```

In fact, this handy shortcut also works with other operators, so you can also write '`n -= 3`', '`half /= 2`', or '`prod *= n`' (the latter will come in handy for exercise 4.3). And, because programmers are inherently lazy, there's even a shorter shortcut for the oft-appearing phrase '`n += 1`', namely '`n++`', meaning 'increase the value of `n` by one'. Similarly, there's '`n--`', which decreases the value of `n` by one. There is no '`n**`' to multiply the value of `n` by one (and it is rarely missed).

Now, since a big percentage of loops occurring in programs have the following form:

```
initialize some loop counter (a)
while ( the loop counter fulfills some condition (b))
{
    /* do something useful here */
    increase the loop counter (c)
}
```

there is another shortcut in C for this type of '*a-b-c*'-loops. This shortcut has the weirdest syntax of any C construct:

```
for (a; b; c)
{
    /* do something useful here */
}
```

Yes, those really are semicolons in there. So, our Gaussian example in the form of a `for`-loop would look like this:

```
int n;
int sum = 0;
for (n = 1; n <= 100; n++)
{
    sum += n;
}
```

or, since a single-statement block can be replaced with the statement itself:

```
int n;
int sum = 0;
for (n = 1; n <= 100; n++)
    sum += n;
```

Compact and cryptic, just the way we C-programmers like it!

Since the variable used as loop counter is usually only interesting in the loop itself, C99 added the feature of allowing to *declare* the loop counter inside the `for`-construct:

```
int sum = 0;
for (int n = 0; n <= 100; n++)
    sum += n;
```

The variable `n` is then only valid inside the body of the loop. This feature was 'borrowed back' from C++.

When you make a `while`-loop and the loop condition doesn't evaluate to `true` to begin with, the loop is simply skipped:

```
int n = 0;
while (n > 0)
{
    /* code in here is never executed */
}
```

There are cases when you'd like the loop to be run at least once. Consider the following program, which reads numbers as you type them (separated either by spaces or with 'Enter') and adds them, until you type a zero at which it prints out the sum and exits:

```
#include <stdio.h>

int main(void)
{
    int sum = 0;
    int n;
    scanf("%i", &n);        /* get the first number */
    while (n != 0)          /* only continue if nonzero */
    {
        sum += n;
        scanf("%i", &n);    /* get the next number */
    }
```

```
    printf("sum = %i\n", sum);
    return 0;
}
```

(It may be a good idea to actually type in and compile this program, to see how it works.) By the way, the `scanf` lines may still look a little strange to you – you probably recognize the `%i` from `printf`, but I promise we'll get back to the strange ampersand before `n` later.

You also notice that there are two identical lines in the program: the '`scanf`' appears twice. This is almost always a bad sign. In this case, it's just a single statement, but what if there were complicated calculations involved in getting the next `n`? You would have to type them both before starting the loop *and* inside the loop. For these cases, there's a third kind of loop in C: the `do-while`-loop. In this loop, the loop condition is tested *after* the first iteration, so the program would look like this:

```
#include <stdio.h>

int main (void)
{
    int sum = 0;
    int n;
    do
    {
        scanf("%i", &n);
        sum += n;
    } while (n != 0);
    printf("sum = %i\n", sum);

    return 0;
}
```

There is something which needs to be said about `scanf`. It reads from the 'standard input' (usually the keyboard, but we will see later that this can also be a file) looking for data in the correct format (as specified in the format string). As soon as it finds a character which doesn't 'fit', it stops. The non-matching character will be the first one read in the next `scanf()`.

This means you can enter numbers on separate lines, or type several of them separated by white space on a single line. It also means it is rather sensitive to the input given. Especially inside a loop, as in the program above, this can lead to problems. If the user of this program enters anything which is not a digit (perhaps typing 'quit' because he doesn't know to end the program by entering a zero), this will lead to an unpleasant surprise, which we will cover in the next section.

4.5 Caveats

Loops are a very powerful language feature, and it is hard to come up with an example of a 'real-world' program which doesn't use them. However, they introduce a whole new class of errors which can cause no end of trouble. Quite literally. Consider the following program:

```
int sum = 0;
int n = 1;
while (n <= 100)
{
    sum = sum + n;
}
printf("The sum of 1 through 100 is %i\n", sum);
```

As you can see, we 'forgot' the line 'n++;' in the loop. The compiler will think nothing of it, but can you imagine what would happen when you'd run this program?

You could wait until the cows come home, but you would never see the line 'The sum of 1 through 100 is 5050' appear. It may look like the computer is just sitting there doing nothing, but there's something else going on. If you are on a system which lets you see the 'CPU load' (in Windows, you can look at the CPU meter in the Task Manager, and on UNIX systems you could type uptime), you can see that the computer is actually quite busy.

So, what's going on? The computer enters the loop (since n = 1 and thus n <= 100 holds), adds n to sum, then checks whether n <= 100 (which it is, since it hasn't changed!), after which is adds n to sum again, etc., etc. In other words, it never exits the loop.

This is a classic example of an *infinite loop*[2]. The only way to 'break' out of it is by using 'brute force'. On most systems, you can interrupt a running program by typing ^C (that's Control-C).

The problem, though, is how do you *know* your program has entered an infinite loop? Since computers are not infinitely fast, any program will take a finite time to complete. Especially if you will later write complex programs which do lots of hefty calculations, it may *look* like the computer is stuck in an endless loop, when it is really in the middle of some lengthy calculation.

The idea is that your program should periodically inform the user that it is still working. Users (including yourself, when you run your own program) don't particularly like it if they have to stare at a blank screen for seven minutes after starting the program, after which it'll print out 'Thanks for your patience, the answer is 42' or something like that.

Instead, you should print out what your program is doing. For example, if an algorithm involves looping ten times over a calculation which takes thirty seconds, it is nice to print out that it is about to start calculation n of 10. This not only tells the user that the program is actually still running, but also provides an estimate about how much longer (s)he needs to wait for the answer.

Okay, you are probably convinced by now that infinite loops are a bad thing, and that you will never write one. However, see if you can spot the problem in the following excerpt:

[2]If you look up *infinite loop* in a computer-savvy dictionary, it often includes '*see:* infinite loop' in its description, which sums it up pretty nicely.

```
int sum = 0;
int n = 1;
while (n <= 100);
{
    sum = sum + n;
    n++;
}
printf("The sum of 1 through 100 is %i\n", sum);
```

If you can't see it straight away, one extra hint is that this program won't ever get past the third line. The problem lies in the semicolon ending it. It may be innocuous to a beginning C programmer, but to an experienced programmer, such a semicolon sticks out like a sore thumb (just like an `if`-statement with a single '='). In effect, the semicolon ends the whole `while`-loop. So the body of the loop consists of 'nothing'. And the fact that a new statement block is started immediately after the `while`-line is no problem to the C compiler, since you are free to start a new block in the middle of another one.

A similar problem can occur with `if`-statements:

```
int a = 3;
if (a < 2);
    printf("Strange!\n");
```

Here, indenting suggests that the `printf`-statement is within the scope of the `if`-statement, but that's just misleading. Just like in the example above, the semicolon after the `if`-statement effectively ends it, and the program happily continues with the next line. And again, the compiler needn't find anything wrong with it. Had you typed

```
int a = 3;
if (a < 2);
    printf("Strange!\n");
else
    printf("All's well...\n");
```

then the compiler would have caught it, since then there'd be an `else` statement outside of an `if` statement, and that doesn't make sense. The compiler may not directly say so, though; it may also say something like 'parse error before 'else''. Slightly confusing since the actual problem is located earlier in your code.

At the end of the previous section, an 'unpleasant surprise' was mentioned in the following little program:

```
#include <stdio.h>

int main (void)
{
    int sum = 0;
    int n;
```

```
        do
        {
            scanf("%i", &n);
            sum += n;
        } while (n != 0);
        printf("sum = %i\n", sum);

        return 0;
    }
```

If you run this and type a non-digit as one of its inputs, this will result in an infinite loop as well. The reason for this one is quite hard to spot. It is because of the feature of the `scanf()` function that as soon as it finds a non-matching character, it stops *and the non-matching character will be the first one read in the next call to* `scanf`. However, the next `scanf()` in this case will *also* be looking for a number, so the character *again* won't match, and `n` will keep its previous value. In other words, nothing is ever 'consumed' from the input anymore, and the program will keep hiccuping at this non-matching character forever.

Fortunately, `scanf()` returns the number of items succesfully scanned, and we can get around this problem by checking this number:

```
do
{
    if (scanf("%i", &n) != 1)   /* if we can't scan the input... */
        n = 0;                  /* ...force it to zero */
    sum += n;
} while (n != 0);
```

With this modification, you can end the program by typing any non-digit as input.

4.6 The Ternary Operator

A cool little feature of C is the so-called *ternary operator*, which has the form

boolean expression ? *true-part* : *false-part*

The result of this expression is the *true-part* if the boolean expression in front of the question mark evaluates to *true*, or else the *false-part*. The power of this operator shows when we re-write the function we started this chapter with:

```
int abs(int x)
{
    if (x < 0)
        return -x;
    else
        return x;
}
```

becomes

```
int abs(int x)
{
    return x < 0 ? -x : x;
}
```

This ternary operator is often used if you have a function call with lots of parameters and only one of them is dependent on a certain boolean expression. Instead of typing this function call twice inside an `if`-statement, with only one of the function parameters differing, you can type it once with this one parameter replaced by a ternary expression.

Another very common use of it is in combination with macros (see 3.11):

```
#define MIN(a, b) ((a) < (b) ? (a) : (b))
#define MAX(a, b) ((a) > (b) ? (a) : (b))
```

which return the minimum of the two values passed to it, or the maximum, respectively. In fact, the whole `abs()` function could be replaced with a similar macro:

```
#define ABS(x) ((x) < 0 ? -(x) : (x))
```

4.7 Scope

Every time you open a new block with an 'opening curly brace' {, you introduce a new *scope*. Simply said, a scope is 'where variables live'. After you have declared a variable, you can use that variable throughout its scope. We have briefly mentioned the fact that variable declarations can only appear at the start of a scope in the previous chapter (a restriction which was removed in C99), but we have now seen that new scopes can appear within other scopes. And in each new scope, you can declare new variables, so the following is perfectly legal even in ANSI C:

```
#include <stdio.h>

int main(void)
{
    /* This is the scope of the main function */
    int a = 1;
    if (a > 0)
    {
        /* This is a new, 'inner' scope, so we can
           declare a new variable here */
        int b = 2*a;
        printf("b = %i\n", b);
    }
    return 0;
}
```

Note that variables from one scope are also available in any of its inner scopes. This is why the variable a could be used to initialize b in the example above.

More specifically, a variable is *only* visible in its own scope and any inner scope. So, the following would not work:

```
int main(void)
{
    int a = 1;
    if (a > 0)
    {
        /* a is known here */
        int b = 2*a;
        printf("b = %i\n", b);
    }
    /* we have left the inner scope, so b is no longer known */
    printf("b = %i\n", b);
    return 0;
}
```

since b was declared in the inner scope and is 'destroyed' as soon as your program leaves that scope. The compiler will flag this as an error.

Although we have so far only seen scopes in if-statements or as the body of loops, you are actually free to create a new scope whenever you please:

```
int main(void)
{
    int a = 3;
    printf("a = %i\n", a);

    /* we may not declare a new variable here anymore... */
    {
        /* ...but in this new scope, we can! */
        int b = 2*a;
        printf("b = %i\n", b);
    }
    /* we have left the inner scope, so b is no longer known */
    return 0;
}
```

What goes on 'behind the scenes' is that whenever a new variable is declared in a scope, enough memory to hold it is reserved in a special area of the system memory called the *stack*. Things allocated on the stack can be thought of as notes on a stack of paper: whenever room is required to write down something new, a new note is placed on top of the stack. When it is no longer needed (*i.e.*, at the closing curly brace of the scope), the piece of paper is removed. This operates in a strict 'last in, first out' order: the last thing to be allocated on the stack is the first thing to be removed from it (and is henceforth no longer available).

An interesting thing happens when you declare a new variable in an inner scope which is already used in the outer scope:

```
int main(void)
{
    int a = 3;
    printf("a = %i\n", a);
    {
        int a = 4;
        printf("a = %i\n", a);
    }
    printf("a = %i\n", a);
    return 0;
}
```

Try it, and see what the program prints out. You may be surprised to see that the value of `a` in the outer scope is not changed by what happens in the inner scope. This is because the 'inner' `a` is not the *same* `a` – it is a new variable which just happens to have the same name. It is said that this inner variable 'shadows' the outer one. Some compilers may warn you when this happens, but it *is* perfectly valid C.

Something similar happens when you implement a function:

```
int f(int a)
{
    int a = 2;
    return a;
}
```

When you compile a program containing this function, you probably get a warning along the lines of `warning: declaration of 'a' shadows a parameter.` Many compilers treat it as an error. If you are *really* adventurous, try this:

```
int f(int a)
{
    int a = 2*a;
    return a;
}
```

Try calling this function and printing out its result. Even if your particular compiler lets you get away with this, don't ever use it in real code, of course.

There is also one 'global scope'; if you declare a variable outside of any functions, it is available throughout all other scopes, just as if your entire file started with a single `{` and ended with a single `}`. It is considered 'bad style' to put variables in the global scope, and it is only necessary in exceptional cases.

Incidentally, you are free to 'nest' scopes, just like you could add multiple parentheses. So this is fine:

```
int f(int a)
{
    {
        {{{ return a*a; }}}
    }
}
```

Also, you can nest loops and if-statements several levels deep:

```
int main(void)
{
    int i;
    int j = 0;
    while (j < 42)
    {
        for (i = 3; i < 17; i++)
        {
            if (j < i)
            {
                printf("%i < %i\n", j, i);
            }
            else
            {
                printf("%i >= %i\n", j, i);
            }
        }
    }
    return 0;
}
```

(Warning: This program generates quite a lot of output!) Incidentally, apart from the outer curly braces, you can leave all the braces out, since a loop + single statement in itself counts as a single statement, as does a complete if-construct:

```
int main(void)
{
    int i;
    int j = 0;
    while (j < 42)
        for (i = 3; i < 17; i++)
            if (j < i)
                printf("%i < %i\n", j, i);
            else
                printf("%i >= %i\n", j, i);
    return 0;
}
```

This is definitely less readable, so adding curly braces anyway is a good idea.

4.8 A Useful Example: Roots

In many problems in science, an important step is finding the *root* of a certain function, *i.e.* the value x for which $f(x) = 0$. Of course, for many functions this can be solved by using straight-forward algebra, but sometimes it is difficult or impossible to find the answer algebraically. In such cases, we have to settle for an *approximation* to the answer, which we obtain *numerically*. There are many different ways of doing this, and whole books have been written about the subject. We will focus on a very simple algorithm here, known as *bisection*.

First, we will formulate the problem in a slightly more formal way: Given a function $f(x)$ which is known to have a single root on the interval $[a, b]$ (*i.e.* there is one (and only one) $x_0 \in [a, b]$ such that $f(x_0) = 0$), find x_0 to within ε (or more precise: find x_1 such that $x_1 - \varepsilon < x_0 < x_1 + \varepsilon$).

The first go at a program to solve this problem might implement the following algorithm: Simply start walking from a to b in steps of ε, see if the function at the current point evaluates to zero, and print out the current point if so (assuming an interval of $[0, 1]$ and an ε of 10^{-6}:

```
#include <stdio.h>

double f(double x)
{
    /* some function */
}

int main(void)
{
    double a = 0;
    double b = 1;
    double epsilon = 1e-6;

    double x = a;
    while (f(x) != 0 && x < b)
        x += epsilon;

    if (x >= b)
        printf("No root in the interval!\n");
    else
        printf("Found x0 = %f\n", x);

    return 0;
}
```

For many cases, this program will not find a root even if it lies within the specified interval. This is because we explicitly check for zero, which we saw earlier is a dangerous thing to do in floating point calculus. If $f(x)$ evaluates to 'slightly above zero' yet $f(x + \varepsilon)$ evaluates to 'slightly below zero', we know that the root lies within $[x, x + \varepsilon]$, but the program above misses it.

One approach would be to check whether $f(x)$ is 'within δ' of zero (for some small value of δ), but this is dangerous too: If the function under investigation 'moves close to zero' but doesn't quite reach it, the 'within δ' approach may falsely report a root.

A more 'robust' approach is to check for 'zero crossings': if $f(x) > 0$ and $f(x + \varepsilon) < 0$ (or vice versa), then the root must be in the interval $[x, x + \varepsilon]$:

```
while ((f(x) < 0 && f(x + epsilon) < 0)
    || (f(x) > 0 && f(x + epsilon) > 0)
    && x < b)
    x += epsilon;
```

To test our program, let us fill in a simple `cos(x)` for `f(x)`, and adjust the upper limit (b) to 2. We know there's a root at $x_0 = \frac{\pi}{2}$, so let's see if the program finds it. We'll also modify the program to print out how long it took to find the root, and the error in the approximation (because we happen to know the exact value).

```
#include <math.h>
#include <stdio.h>
#include <time.h>

#define HALF_PI 1.57079632679489661923

double f(double x)
{
    return cos(x);
}

int main(void)
{
    double a = 0;
    double b = 2;
    double epsilon = 1e-6;
    double x = a;

    clock_t end;
    clock_t start = clock();

    while ((f(x) > 0 && f(x + epsilon) > 0)
        || (f(x) < 0 && f(x + epsilon) < 0)
        && x < b)
        x += epsilon;

    end = clock();
    printf("Found: %f\n", x);
    printf("Error: %g\n", HALF_PI - x);
    printf("Took: %g seconds\n",
        (double)(end - start)/CLOCKS_PER_SEC);
    return 0;
}
```

The `clock()` function returns the number of 'ticks' which have elapsed since the system started up. Each platform is free to define how many of these clock ticks happen per second (usually at least one thousand). The `CLOCKS_PER_SEC` is a `#define` which gives this number. Since these are all integers, we cast the elapsed time to a `double` before doing the division. We `#include` the `time.h` header file which contains all this stuff.

The `%g` format specifiers in the `printf()` calls print out the floating point value using 'scientific notation'. We have to use this because otherwise the error would be printed as `0.00000` – not very useful.

On a Pentium4 system running at 2.4 GHz (considered a moderate system at the time of writing) this program took roughly half a second to find the correct value, with an error of about 3.1×10^{-7} (which is indeed within the given ε). So far, so good!

However, the algorithm we use is not very 'smart'. To demonstrate this, let's stress the program a bit more and ask for one extra digit of precision, by setting $\varepsilon = 10^{-7}$ instead of 10^{-6}. We'll notice that the program now takes about 5 seconds to complete. Still not really an eternity, but definitely noticeable.

4.8.1 Speeding Things Up

One thing which you may have noticed is that for each iteration, the function is evaluated up to 4 times. For a simple `cos(x)` this is not so much a problem, but if we had used a complex function there instead, the time it takes quickly adds up.

Since the function evaluations are actually happening at only two places per iteration (namely, x and $x + \varepsilon$) and since the 'right value' becomes the 'left value' at the next iteration, we can improve a bit by 'saving' the results:

```
int main(void)
{
    double a = 0;
    double b = 2;
    double epsilon = 1e-7;
    double x = a;
    double fx;   // for f(x)
    double fxe;  // for f(x + epsilon)

    clock_t end;
    clock_t start = clock();

    fx = f(x);
    fxe = f(x + epsilon);

    while ((fx < 0 && fxe < 0) || (fx > 0 && fxe > 0) && x < b)
    {
        x += epsilon;
        fx = fxe;     /* re-use the previous value */
        fxe = f(x + epsilon);
```

```
    }

    end = clock();
    printf("Found: %f\n", x);
    printf("Error: %g\n", HALF_PI - x);
    printf("Took: %g seconds\n",
        (double)(end - start)/CLOCKS_PER_SEC);
    return 0;
}
```

With this modification, the program ran in slightly over 2 seconds on the same system – an improvement of a factor 2.5 with only a little effort.

4.8.2 The Order of an Algorithm

However, the algorithm still isn't very good. Each time you make ε a factor of ten smaller (which you need to do to gain one extra digit of precision), the program will take ten times as long to reach the answer. If we wanted to use the current program to get, say, *ten* digits of precision instead of seven, it would take a thousand times as long – on the 2.4 GHz Pentium4 used in our tests, that would amount to over half an hour. Going to twelve digits would take two full days!

An algorithm like this is said to be $\mathcal{O}(10^n)$. This so-called 'big-O notation' means that the *order* of the algorithm is 10^n, (*i.e.* the algorithm scales with 10^n), where n, in this case, is the number of significant digits in the result. $\mathcal{O}(10^n)$ is about as bad as it gets. Luckily, there are quite a few better alternatives for finding the root of a function in a given interval, and we'll look at an $\mathcal{O}(n)$ one next: each time you halve ε, the algorithm takes just one extra iteration.

The algorithm we'll look at uses a trick known as *bisection*, and it deserves special attention because its main premise is useful in a variety of problems. In each iteration the 'solution space' is divided in two, and it is determined in which half the solution lies. The other half is simply discarded, and the correct half becomes the new interval. The implementation may look like this:

```
int main(void)
{
    double a = 0;
    double b = 2;
    double epsilon = 1e-7;

    clock_t end;
    clock_t start = clock();

    double fa = f(a);
    double fb = f(b);

    /* value in the middle of the interval */
    double fm = f((a + b)/2);
```

```
    while (b - a > epsilon)
    {
        if ((fa < 0 && fm < 0) || (fa > 0 && fm > 0))
        {
            /* the root lies in the interval [m, b] */
            a = (a + b)/2;  /* adjust the new lower bound */
            fa = fm;        /* re-use the function value */
        }
        else
        {
            /* the root lies in the interval [a, m] */
            b = (a + b)/2;  /* adjust the new upper bound */
            fb = fm;        /* re-use the function value */
        }
        /* one new calculation per iteration: */
        fm = f((a + b)/2);
    }

    end = clock();
    printf("Found: %f\n", (a + b)/2);
    printf("Error: %g\n", HALF_PI - (a + b)/2);
    printf("Took: %g seconds\n",
        (double)(end - start)/CLOCKS_PER_SEC);
    return 0;
}
```

On the same computer system which took 2 seconds to approximate the root to within $\varepsilon = 10^{-7}$ with the 'brute-force' method of the previous subsection, the program above took just 170 μs (that's over ten thousand times as fast)! And the power of the bisection method really shines when we set ε to a *really* low value, like 10^{-15}. The program terminates in about 185 μs, where the 'brute-force' method would have taken about six years...

4.9 Recursion

You are free to call functions from within other functions, but what would happen if you called a function from within *itself*? Consider the following code:

```
void recurse()
{
    recurse();
}
```

When you call this function in a program, you may expect an 'infinite loop' to happen, so that your program will become 'stuck' – just like the example on page 80. In reality, your program will be stuck for a while, but then crash.

The reason is that each time a function is called, what happens 'behind the scenes' is that the computer records where the program should continue after finishing the function (the *return address*), by storing this address on the *stack* (see page 84) – the same memory where local variables are stored. This explains nicely how 'nested function calls' work: Each time a function call is made, the current return address is pushed onto the stack. Whenever a function returns, the system simply looks at the top return address on the stack and continues executing the program there after 'popping' the address from the stack. If the call was made from within another function, that means that there's another return address on the stack, etc.

In the example above, the computer stores the program location when the `recurse()` function is called, then enters the function, and sees another function call. It stores the program location again, enters the same function again, stores the location again, etc.

The difference with an infinite loop is that with 'infinite recursion', these stored return addresses take up memory space on the stack, and stack space is not infinite. Therefore, when the available memory fills up (which may take a while, depending on the speed/memory ratio of your system), there is no more room to store yet another return address, and the program is terminated.

Recursions, of course, normally are not infinite. Consider the following function:

```
int recurse(int n)
{
    if (n <= 0)
        return 0;
    else
        return recurse(n - 1);
}
```

What will be the return value of this function? The answer is that it is *always zero*, regardless of the value you passed into it. If that value was zero or less, it will return zero immediately. Otherwise, it will recurse with a value one less than what you called it with. If that is zero (*i.e.*, you called it originally with a value of one), the recursion is ended. If it isn't (yet), it recurses *again*, with a value one less than the current value, *two* less than your original value. This will go on until it reaches zero.

Now, this is not the *whole* truth. Again, storing return addresses costs memory, and if you pass a really large value to the above `recurse()` function, it *will* run out of memory and crash.

4.10 Synopsis

The *flow* of a program can be more complex than a simple top-to-bottom execution of all the statements in the program. Function calls can be considered to modify the program flow; you can also have *conditional branches* (*i.e.*, parts of the program which get executed if a certain condition is met, and other parts if it is not) and *loops* (*i.e.*, parts of the program which get executed repeatedly until some condition is met).

Conditional branches are made in C with `if` (and optionally `else`) statements or with `switch/case` constructs; loops are expressed using `for`, `while`, or `do-while` constructs.

Boolean expressions are expressions which evaluate to one of two possible 'values': 'true' or 'false'. 'False', in C, is expressed as zero; everything non-zero is considered to be 'true'.

The *scope* of a name (most notably the name of a variable) is that part of the program in which that name is known. For instance, the scope of a local variable, declared at the top of a certain function, is that function (*i.e.* from its declaration up to the final `}` closing the definition of that function). A variable cannot be referred to outside of its scope. Variables in different scopes have nothing to do with each other and can have the same name without interfering.

Calling a function from within itself (either directly or via another function) is called *recursion*.

4.11 Other Languages

Conditional branches and loops are part of most, if not all, programming languages. In some languages, a 'branch' is actually a 'jump' – *i.e.*, a special command which transfers the 'current position' to a different place in the program. C actually has such a command (`goto`), but its use is 'considered harmful'. On a lower level (assembly language), a loop is usually not available as a built-in construct, but any loop can be constructed with a conditional branch/jump.

Since any loop can also be rewritten as a form of recursion, some languages offer recursion as the only way of constructing loops. Lisp is an example of such a language. Other languages, such as FORTRAN77, do not allow recursion at all.

Many languages have explicit support for boolean expressions by offering a special `bool` (C++) or `Boolean` (Pascal) type, along with predefined `true` and `false` values.

C++ and C99 remove the restriction that you can only introduce new variables at the beginning of a new block; Pascal on the other hand wants you to list all variables used in a function in a special section at the top of that function. Some (scripting) languages don't require you to declare variables before you use them and/or give them an implicit value (zero, usually) when they are first encountered. This has the drawback of typos not being caught at compile time, but yielding unexpected results at runtime because they are (silently) replaced with zeros. Some languages don't make a big fuss about scoping either, or allow you to 'export' variables to a larger scope (C allows this too, which we will see more about in chapter 10).

4.12 Questions and Exercises

4.1 Write a `signum()` function which takes a `double` as its single parameter and which returns the *signum* $S(x)$ of that parameter:

$$S(x) = \begin{cases} -1 & \text{if } x < 0 \\ 0 & \text{if } x = 0 \\ +1 & \text{if } x > 0 \end{cases}$$

4.2 Write an `approx_equal()` function which takes three `double`s, `a`, `b`, and `f`, and returns whether `a` is within a fraction `f` of `b`.

4.3 ⋆ Write a `fac()` function which takes an `int` and returns its *factorial*, which is defined as

$$n! = n \cdot (n-1) \cdot (n-2) \cdots 3 \cdot 2 \cdot 1$$

but do not use `for` or `while`. Hint: $n! \equiv n \cdot (n-1)!$.

4.4 Find out the maximum depth of recursion on your system, by taking the example `recurse(int n)` function on page 92 and calling it with larger and larger values of `n`. Then, modify it to take some extra (dummy) parameters. Does the maximum depth of recursion depend on the number of parameters? If so, why do you think that is?

4.5 ⋆ In the final, shortest version of the function to implement the truth table on page 74, the two boolean parameters were compared for equality, which was a bad idea because would break down when calling it with 'true' values other than one. However, with a simple change, you *can* write that function very compact and equivalent with the other example implementations, even lacking a 'real' boolean type, using only a single comparison operator. How?

Chapter 5
Arrays, Pointers, and Memory

It's a poor sort of memory that only works backward.

Lewis Carroll

5.1 Arrays

Suppose you need to calculate statistical information about a series of values. For instance, you want to know the average value, the median, and the standard deviation. And further suppose that you know the number of values beforehand – say 10. You could write a function like this:

```
double average(double v0, double v1, double v2, double v3, double v4,
               double v5, double v6, double v7, double v8, double v9)
{
    double sum = v0;
    sum += v1;
    sum += v2;
    sum += v3;
    sum += v4;
    sum += v5;
    sum += v6;
    sum += v7;
    sum += v8;
    sum += v9;
    return sum/10;
}
```

Surely, there must be a better way. After all, in ordinary math we can write it much more compact:

$$\langle v \rangle = \frac{1}{10} \sum_{n=0}^{n=9} v_n$$

Couldn't we do this in some kind of loop? Something like

```
double average(double v0, double v1, double v2, double v3, double v4,
                   double v5, double v6, double v7, double v8, double v9)
{
     int n;
     double sum = 0;
     for (n = 0; n < 10; n++)
          sum += v_n;
     return sum/10;
}
```

Unfortunately, this won't compile; `v_n` is not turned into some kind of v_n in a magic way. Fortunately, we *can* define vectors of values in C using *arrays*. The function above could be written like this:

```
double average(double v[10])
{
     int n;
     double sum = 0;
     for (n = 0; n < 10; n++)
          sum += v[n];
     return sum/10;
}
```

An array in C is declared like this:

type *name*[*size*] ;

and a particular element of the array is referenced by putting its index in the brackets. Note that C, like grown-ups, starts counting at zero. So arrays start at index zero, hence the array `double v[10]` actually goes from `v[0]` to `v[9]` and the third element is referenced as `v[2]`.

Although the function above was declared as taking `double v[10]`, we might as well have written `double[42]` or `double[1]` or even `double[]`, which is actually the preferred form. The thing is that C doesn't know (or care) exactly how big the array is – just that there's going to be an array of `doubles`. This means we can now also make a more generic function which doesn't need to be changed for a different number of values:

```
double average(double v[], int num)
{
     int n;
     double sum = 0;
     for (n = 0; n < num; n++)
          sum += v[n];
     return sum/num;
}
```

Filling an array and calling the function above is straightforward too:

```
int main(void)
{
    double values[3];
    double avg;

    values[0] = 3.14;
    values[1] = 2.72;
    values[2] = 1.41;

    avg = average(values, 3);
    printf("Average = %f\n", avg);
    return 0;
}
```

When *declaring* the array, you *do* have to specify its size, which needs to be known at compile time. If the values to be put in the array are also known at compile time, you can use a special short-cut to fill the array:

```
int main(void)
{
    double values[] = { 3.14, 2.72, 1.41 };
    double avg = average(values, 3);
    printf("Average = %f\n", avg);
    return 0;
}
```

Note that in this case, you don't need to specify the size of the array between the brackets; the compiler can count just fine. This shortcut will not work if the values cannot be determined at compile time, e.g. because they need to be provided by the user when the program runs. In that case, you will have to loop over the array to fill it yourself.

5.2 Bounds

So far, almost every new concept that was introduced in this book came with a warning about a new type of bugs you could introduce using it. Arrays are no exception, and in fact they give rise to one of the most dangerous and hard to find type of bugs: out-of-bounds referencing.

Consider the following piece of code:

```
int main(void)
{
    double v[] = { 3.14, 2.78, 1.41 };
    double avg = average(v, 4);    /* should have been 3! */
    printf("Average = %f\n", avg);
    return 0;
}
```

What will happen is that in the `average()` function, the code will add up the elements `v[0]`, `v[1]`, `v[2]`, and `v[3]` – which does not exist! This is an out-of-bounds reference – the computer tries to read one past the actual dimension of the array. Your program may crash at this point (which is unlikely in this case) or retrieve whatever value happened to be in memory at this particular spot (most likely some other variable), and treat it like it was a `double`.

What's even more fun is *writing* out-of-bounds:

```
int main(void)
{
    int v[3];
    int i;
    for (i = 1; i <= 3; i++)
        v[i] = i;
    return 0;
}
```

Typical code for people who aren't used to starting counting at zero. What happens is that `v[0]` is skipped; `v[1]` becomes 1, `v[2]` becomes 2, and finally the computer tries to write the value 3 to `v[3]`, which again does not exist. Instead, the computer will poke that value 3 right after the memory allocated for the array – and who knows what *really* lives at that location. This situation is more likely to crash, but it could also lead to other, more subtle bugs, where the value of some other, totally unrelated variable (which just happened to be in memory adjacent to your array) gets overwritten, leading to incorrect program behavior elsewhere.

Some other programming languages do 'bounds checking' for you; *i.e.* every time an array element is referenced, the computer checks whether you are indeed within the bounds of the array, and aborts the operation if not. This, however, goes against the C philosophy: After all, this check does not come for 'free' (in terms of computer performance), and why would every programmer have to pay for the mistakes of a few?

5.3 Strings

In previous chapters, we have encountered the concept of *strings* a few times, and they were defined as 'arrays of `char`s' with a promise of a more elaborate explanation at a later time. That time is now.

String literals (*i.e.*, strings which are known at compile time) can be designated by typing their contents in quotation marks, like so:

```
char mystring[] = "Hello, world!";
```

Again, there is no need to fill in its length in this case, since the compiler can count the characters for you. In fact, you probably would have entered 13 in the brackets above, right? We'll soon see why that is wrong.

5.3.1 Encoding

First, let's examine the array of characters more closely. Remember that in principle, a computer doesn't know what 'characters' (as in 'human readable tokens') are. It just knows about numbers, and in fact, that's exactly what the string contains. The trick is to have an agreement on what number should be printed out as what character.

The most widely used 'agreement' is the 'American Standard Code for Information Interchange', or ASCII for short. In that code, the number 65 is to be interpreted as the character 'A', number 66 as the character 'B', etc.; the number 97 is the lower case 'a', 98 the lower case 'b', etc.; the number 32 is a space, the number 48 is the digit '0', 49 the digit '1', etc. Everybody pretty much agrees on what the numbers 32 – 127 are supposed to mean; numbers below 32 don't always do the same on different machines (they are so-called 'control codes' for repositioning the current position to the beginning of the line, to use the tabulator, to sound a beep, etc.) and the numbers 128 through 255 contain things like accented letters or extra symbols. The problem is that in the past, each computer vendor put its own set of symbols there.

If you're interested in what the actual codes are, it's very easy to find out:

```
#include <stdio.h>
#include <string.h>

int main(void)
{
    char mystring[] = "Hello, world!";
    int len = strlen(mystring);
    int i;

    for (i = 0; i < len; i++)
        printf("%c = %d\n", mystring[i], mystring[i]);

    return 0;
}
```

The function `strlen()` takes a string and returns the number of characters in it. It's declared in `string.h`, which is why we `#include` that. The interesting part is that we print out each character in the string both as the character (`%c`) and as a number (`%d`). You see that a string can be indexed like any other array.

Incidentally, the 255 possible character codes are not enough to cover all possible symbols and characters you may want to use on your computer. Most computer vendors have put at least part of the Greek alphabet in there, for instance, but if you happen to need a particular letter from a different alphabet (like Hebrew, Arabic, or even Chinese) you're out of luck.

There have been some initiatives to make computer text encodings more international, for instance by going to 16-bit 'characters' (giving 65536 possible character codes, in the 'Unicode' encoding), of which the drawback is that it is incompatible with existing code, and that it takes twice as much memory to store strings, even if you don't use any 'strange' characters anyway. Another way is to have ordinary `char`s for your encoding, but agree on certain 'special' codes which mean 'this is not an actual character, but together with the next byte (or next n bytes) forms a special character' (the UTF-8 encoding, which is used a lot on the Internet, uses this).

Since we will be primarily concerned with calculations and numbers, we will forget about multi-byte characters for the remainder of this book.

5.3.2 Termination

Consider a function which takes such a string and prints it. As luck would have it, we already know such a function: `printf`. We can call it to print out our string above like this:

```
printf(mystring);
```

Do you notice anything different from our function call to the `average` function, which took an arbitrarily sized array of `doubles`? If not, try to figure out how `printf` knows how long your string is, since you do not pass its length as an extra parameter. In other words, why don't you have to do

```
printf(mystring, 13);
```

or something like that? Also, *how* was the `strlen()` function in the previous section able to determine the number of characters in the string? The answer is that in C, strings are not just 'arrays of characters'. They have a special feature, namely that the last character of the array is zero. What you often hear is that 'C strings are zero-terminated'. The compiler takes care of this for you, so when you type

```
char mystring[] = "Hello, world!";
```

this is actually equivalent to

```
char mystring[] = { 72, 101, 108, 108, 111, 44, 32, 119, 111,
                    114, 108, 100, 33, 0 };
```

(but obviously, more readable). This is the reason that our example string doesn't use 13, but *14* characters.

5.3.3 String Manipulation

C is not really the language of choice when it comes to string manipulation. If for a particular project, you need to do lots of fiddling with strings (concatenation, search/replacing, etc.), you're better off using a different language.

What you *can* do, however, is index individual characters in a string, since it's an array like any other (apart from that terminating zero, that is). Individual character literals are designated by putting them between single quotes, so the character 'a' is designated `'a'`. For example:

```
#include <stdio.h>

int main(void)
{
    char mystring[] = "Epidermal\n";
    mystring[0] = 'S';
    mystring[8] = 'n';
    printf(mystring);
}
```

Try this program, and note how it changes a scientific word into a popular comic hero.

By the way, it is also allowed to initialize a string like so:

```
char *str = "This is my string";
```

but there is an important difference: In the former case, `mystring` is a character array, initialized with the contents of the literal `"Epidermal"`; in the latter case, `str` is a pointer to the array of characters `"This is my string"`. The difference is that in the former case, you can change the contents of the string (like the example does); in the latter case, the string is located in memory which you aren't necessary allowed to modify.

Remember our little program to calculate the sum of a set of numbers entered by the user (see page 79). It was ended by the user entering a zero. This was a reasonable solution, because a zero wouldn't have any effect on the sum anyway. However, suppose we modify the program to calculate the *average* of a set of numbers entered by the user. In that case, any value entered by the user (including zero and negative values) are valid input. We will use strings to make a program that ends when the user enters '`quit`':

```
#include <stdio.h>
#include <string.h>
#include <stdlib.h>

int main(void)
{
    char s[80];    /* 'enough' room for input */
    double sum = 0;
    int n = 0;

    while (1)      /* loop forever */
    {
        scanf("%79s", s);               /* get user input */
        if (!strcmp(s, "quit"))    /* did we get 'quit'? */
            break;       /* in that case, break the loop */
        sum += atof(s);    /* convert to double and add */
        n++;
    }
```

```
        if (n > 0)      /* avoid division by zero */
            printf("average = %f\n", sum/n);

        return 0;
    }
```

First, we allocate enough room for each line of user input. The number 80 in there is just a guess. This number should be 'large enough' to hold any possible user input. You may wonder how large 'large enough' is, and that's a valid concern: you don't know. The '%79s' in the `scanf` means 'scan for a string of at most 79 characters' (we need to leave one byte for the terminating zero). Had we simply used '%s', and the user typed more than one line full of text (on most systems, a standard line of text is 80 characters wide), we would have a problem – our program would probably crash due to writing pas the available memory range. This is as bad as it sounds. There is a lot of software out there in the world which uses this kind of fixed-size buffers, and it is the number one target for hacker attacks. You can often crash a computer remotely simply by sending a 'large enough' string of text to an unsuspecting program which is listening to remote clients. This is the so-called *buffer overflow attack*. We are not *really* concerned with 'safety' of our programs for the time being, but we'll get to safer ways of handling user input later.

Phew, there was a lot to say about that first line, and only three lines further we see a construct which should have you raise an eyebrow: `while (1) ...` You remember that the `while` loop is repeated for as long as the expression is non-zero, and '1' is *always* non-zero. So in effect, this loop will continue forever. We'll see in a minute why we can get away with this.

We have seen the `scanf` statement before. It is like `printf`, but for input instead of output. This line will enable the user to type in some text (ended with the Enter key) which will be put into the string `s`.

The next interesting statement is the `strcmp` ('string compare'). It takes two strings (in this case, the string just entered by the user, and one literal string) and returns zero if they are identical (and non-zero if not). The line could also be written

```
if (strcmp(s, "quit") == 0)
```

but we remember from §4.3 that the boolean expression '`x == 0`' is equivalent with '`!x`'. This is standard idiom for C programmers. Note that you really do need to call `strcmp` to compare strings, and that simply doing

```
if (s == "quit")
```

means something altogether different: it compares the value of the *pointer* `s` to the *location* of the string literal `"quit"`.

So, if we got the string `"quit"`, we want to end the loop. The `break` statement does that, and the program will continue after the loop (even an otherwise endless loop).

If we got something other than `"quit"`, we use the function `atof()` ('ASCII to floating point') to convert the contents of the string to a floating point value (often a `double` rather than a `float`, but that doesn't matter), add it to `sum`, and increment the count of numbers we've added so far. The `atof()` function is declared in `stdlib.h` (for 'standard library'), which we `#include`.

Note that `atof()` will try to convert *anything* thrown at it to a `double`, even if the user entered `"stop"`, for instance. In that case, it will return 0. In fact, it will convert as much as it can, until it finds a character which doesn't make sense. So `atof("1.23hello")` will return 1.23.

By the way, it doesn't hurt to add a short `printf` at the beginning of the program explaining what it does and how to exit, because it is quite annoying for a future user of the program to be confronted with a seemingly never-ending loop, and be left to guess whether the program terminates by entering `exit`, `stop`, `end`, or `bye`.

At the end of the program, we print out the average value by dividing the sum by the number of values entered (making sure we don't divide by zero, in case the user enters `quit` before entering any numbers).

Since strings are not a very important part of the type of programs this book is aiming at, we will leave the subject for now and return to arrays of numbers.

5.4 Multi-Dimensional Arrays

In C, you are not limited to one-dimensional arrays. You can declare matrices of arbitrary dimension. For example, a two-dimensional, three-by-four matrix of `doubles` would be declared like so:

```
double A[3][4];
```

after which A_{ij} is addressable as `A[i][j]`.

Note that it is a common mistake to type `A[i,j]`. There is such a thing as the 'comma operator' in C, so '`i,j`' is a valid expression (it means: calculate both subexpressions and yield the result of the second).

When the contents of the matrix are known at compile-time, you can enter the values in a similar fashion as for arrays:

```
double A[][4] = { {0, 1, 2, 3}, {1, 2, 3, 4}, {2, 3, 4, 5} };
```

I.e., you specify the rows of the matrix like they were three one-dimensional arrays (which, in fact, they are). Note that you can leave off the first dimension in the declaration, but not any other.

There are some caveats here as well. If you make a function which expects a matrix of certain dimension as a parameter and pass it a matrix of a different dimension, it merely yields a *warning* and not an error. The compiler may say something like '`warning: passing arg 1 of 'f' from incompatible pointer type`' when you pass a three-dimensional matrix to a function expecting a two-dimensional one, or passing a seven-by-eight matrix to a function expecting a two-by-three one. We will get to what a pointer is soon. For now, the lesson is to just pay attention to these warnings (and all other warnings, in fact).

5.5 Pointers

Pointers, in computer science, are subject of controversy and are shrouded in mystery. Apparently, there are two kinds of human beings: Those who can handle pointers, and those who can't. If you are to be fluent in C, you should belong to the first group. Fortunately, the concept of pointers is not complicated at all, as long as one doesn't try to explain them while forcedly avoiding their implementation details.

If you remember from §1.2.2, a computer system operates on data which resides somewhere in *memory*. The computer can store and retrieve data because each memory location has an *address*. A *pointer*, in effect, is nothing but such an address. You can find out the address at which a particular variable is stored using the 'address-of' operator `&`:

```
#include <stdio.h>

int main(void)
{
        int a = 42;
        printf("The variable a (%i) resides at %p\n", a, &a);
        return 0;
}
```

The `%p` in the `printf` is used to print out pointers ('memory addresses', for now). The actual memory location where `a` ends up is dependent on your computer architecture and operating system.

To store such a memory address, you use a *pointer variable* which is declared using the `*` symbol:

```
#include <stdio.h>

int main(void)
{
        int a = 42;
        int *p = &a;
        printf("The variable a (%i) resides at %p\n", a, p);
        return 0;
}
```

The bold-faced line reads: '`p` is a pointer-to-an-`int`, and it is assigned the address of the `int` variable `a`'. If you can read this line without blinking your eyes, you are well on your way to becoming a proficient C programmer. The name 'pointer' is understandable if you view `p` as not being 'the real variable `a`' but merely 'pointing at `a`'.

There is also a 'dereferencing operator', which is a fancy word for 'give me the actual value that this pointer points to' (or, in more hardware-related terms: 'give me the contents of this here memory address'), for which the character `*` is 're-used':

```
#include <stdio.h>

int main(void)
{
    int a = 42;
    int *p = &a;
    printf("The variable a is stored at address %p\n", p);
    printf("Its value is %d\n", a);
    printf("Which is equivalent to %d\n", *p);
    return 0;
}
```

In other words: '`*(&a) == a`'.

The interesting part is that you can also *write* through a pointer:

```
#include <stdio.h>

int main(void)
{
    int a = 42;
    int *p = &a;
    printf("The variable a has the value %d\n", a);
    *p = 43;
    printf("And now its value is %d\n", a);
    return 0;
}
```

You will see that the value of a has changed! This can be used to make functions which modify their parameters:

```
void increment(int *p)
{
    *p += 1;
}

int main(void)
{
    int n = 3;
    increment(&n);
    printf("n = %d\n", n);
    return 0;
}
```

The `increment()` function is defined as taking a 'pointer-to-an-`int`', and it increments whatever p points to by one. Calling this function with the address of `n` (*i.e.* with a pointer to `n`) increments the value of `n` by one.

If you recall the very first C program we saw, on page 32), you will notice it used the `scanf()` function. Here is the program again:

```
#include <stdio.h>

int main(void)
{
    float a, b, c;
    printf("Please enter two numbers: ");
    scanf("%f %f", &a, &b);
    c = a + b;
    printf("Their sum is %f\n", c);
    return 0;
}
```

The `scanf()` function is a typical example of a function which modifies its arguments, namely by changing their contents to whatever the user enters on the keyboard. This is the reason we had to prepend those 'address-of' operators.[1]

A similar caveat as with (multi-dimensional) arrays applies here. The following program will compile without errors (but probably with a warning):

```
void increment(double *p)
{
    *p += 1;
}

int main(void)
{
    double a = 3.14;
    int b = 42;
    increment(&a);
    increment(&b);
    printf("a = %f, b = %d\n", a, b);
    return 0;
}
```

The bold-faced line in the program above will probably trigger a warning (something about 'incompatible pointer type') but it is not an *error*, and the program will compile. If you run it, you will probably get strange results. The reason is that the `increment` function expects the memory location pointed to by `p` to contain a `double`, and interprets the bits it finds there as if they formed a `double`. Adding 1 to a `double` does something altogether different to the actual bits than adding 1 to an `int` (see §1.3.2), so the resulting bit pattern interpreted as an `int` again can have a wildly different value from what you expected.

This is the main reason people hold a grudge against pointers: the way they are used in C pokes a hole in the type-safety of the programming language. They are, however, extremely powerful if you know what you're doing – something we have heard before about C.

[1] Strictly, the `scanf` function *doesn't* change its arguments – the arguments are *pointers* to the variables, and these *pointers* aren't changed! In this stricter sense, you can't write a function in C which changes its argument.

You can even write programs which compile without warnings, yet which do some pretty strange things:

```
#include <stdio.h>

int main(void)
{
    int a;
    *((double*)&a) = 3.14;
    printf("a = %d\n", a);
    return 0;
}
```

The bold-faced line combines casting with pointers. It may look to you like somebody held down the shift key on his keyboard and randomly pressed some number keys, but it is in fact valid C. First, the address of the integer variable `a` is taken by prepending the 'address-of' operator. Then, this address is cast to a pointer-to-`double` by prepending `(double*)`. This is a weird thing to do and the compiler may well raise an eyebrow internally, but it will have to assume you know what you're doing. Finally, this 'fake' `double` pointer is dereferenced and the value 3.14 is poked into the memory used to store the variable `a`. In the `printf` on the next line, the contents of that memory are interpreted as an `int` – which looks quite different from the value 3.14.

It is good to remember that just because you *can* do something, it doesn't mean you *should*.

C also has the concept of 'pointer-to-a-function', which is explained in detail in section 10.6.

5.6 Returning Arrays From Functions

We can be very terse about this subject: You can't. Suppose you want to define functions `inner3()` and `outer3()` to return the inner product and outer product of two three-dimensional vectors, you would probably write something like this:

```
double inner3(double a[3], double b[3])
{
    return a[0]*b[0] + a[1]*b[1] + a[2]*b[2];
}

double[3] outer3(double a[3], double b[3])
{
    double result[3];
    result[0] = a[1]*b[2] - a[2]*b[1];
    result[1] = a[2]*b[0] - a[0]*b[2];
    result[2] = a[0]*b[1] - a[1]*b[0];
    return result;
}
```

The inner3() function is just fine (we put all those 3's in there to do as much as possible to help prevent the 'calling-a-function-with-arrays-of-incorrect-dimension' problem), but this outer3() won't compile. The compiler will probably be quite honest about its failure and tell you something about the function outer3 being defined as returning an array.

You can make the function compile by making it return a double* instead (perhaps because you heard someone say that 'pointers and arrays are really the same thing in C'):

```
double *outer3(double a[3], double b[3])
{
    double result[3];
    result[0] = a[1]*b[2] - a[2]*b[1];
    result[1] = a[2]*b[0] - a[0]*b[2];
    result[2] = a[0]*b[1] - a[1]*b[0];
    return result;
}
```

This will 'only' yield a warning or two. One will likely be something like 'return from incompatible pointer type', and the other something like 'function returns address of local variable'. The former is a sign that the compiler thinks that arrays and pointers are *not* quite the same thing, and the latter warns you about something really bad. Suppose we ignore the warnings and try to call our outer3() function like this:

```
int main(void)
{
    double a[3] = {2, 1, 7};
    double b[3] = {3, 1, 4};
    double c[3] = outer3(a, b);
    printf("(%f, %f, %f)\n", c[0], c[1], c[2]);
    return 0;
}
```

Then the compiler will complain more ardently, and tell you that the bold-faced line contains an 'invalid initializer'. You can get rid of this particular error (and exchange it for a different one):

```
int main(void)
{
    double a[3] = {2, 1, 7};
    double b[3] = {3, 1, 4};
    double c[3];
    c = outer3(a, b);
    printf("(%f, %f, %f)\n", c[0], c[1], c[2]);
    return 0;
}
```

This time, the compiler will say something about '`incompatible types in assignment`'. Perhaps we should conclude that arrays and pointers are definitely not the same thing in C, and try a different approach:

```
int main(void)
{
    double a[3] = {2, 1, 7};
    double b[3] = {3, 1, 4};
    double *c = outer3(a, b);
    printf("(%f, %f, %f)\n", c[0], c[1], c[2]);
    return 0;
}
```

This program will compile without errors, but will be totally wrong. The reason for this is that we returned not the actual values in the array (*i.e.*, not the array itself) but rather a pointer to it, *i.e.* the address at which this array resided. Note the use of the past tense here, because as soon as a program exits a function, any locally declared variables are cleaned up (they are literally 'popped from the stack' and the memory they occupied is returned to the system), because a function body is a *scope*. Hence, the pointer points to 'stale' memory. You can try running the program; it will probably print out some *really* weird numbers.

There are ways around the 'cannot-return-arrays-from-functions' limitation in C. One is by explicitly allocating memory in the function and returning a pointer to that memory (which we'll cover in the next section) or by 'wrapping' the array in a `struct` (which we'll cover in the next chapter), another is to use one of the function parameters as a so-called 'out value':

```
void outer3(double a[3], double b[3], double result[3])
{
    result[0] = a[1]*b[2] - a[2]*b[1];
    result[1] = a[2]*b[0] - a[0]*b[2];
    result[2] = a[0]*b[1] - a[1]*b[0];
}
```

This works because the `outer3` function does not receive a *copy* of the arrays, but rather only pointers to their first elements. More on this in the remainder of this chapter and in the next one.

5.7 Dynamic Memory Allocation

The examples so far assumed that you knew beforehand (*i.e.* when typing in your program) how large your arrays were going to be. Suppose, however, that this size is the result of a computation, or is the result of some value entered by the user of your program at run time. Consider the following code excerpt:

```
void f(int n)
{
    int array[n];
    ...
}
```

In 'old' standard C, this is illegal. The size of an array needs to be known at compile time, so the example above won't fly – even if you only call this function with code like

```
f(42);
```

because the compiler doesn't 'foresee' that you're only calling this function with a compile-time constant value.

However, having arrays with sizes determined at run time (sometimes called *variable length arrays* or VLAs) is very useful. Some compilers have allowed this as an 'extension' to the C language (most notably GCC) and this feature has been added to the C99 standard (more on that later in this chapter).

Since it is still illegal to return an array from a function (and because it is useful in other scenarios as well), we will look at two functions supplied by the C standard library to handle dynamic memory allocation: the functions `malloc()` and `free()`. These are declared in the header `stdlib.h`. Their prototypes are

```
void *malloc(size_t num_bytes);
void free(void *ptr);
```

The `malloc()` function returns a pointer to a chunk of memory of (at least) the specified size. On most systems, memory is handed out in multiples of some minimum size (often 8 or 16 bytes). This memory is said to be allocated on the *heap*, as opposed to memory which is allocated on the *stack* automatically for 'normal' variables as they are declared (see page 84). Also, as opposed to stack allocations, there is no 'last in, first out' order. Once you have received a pointer, it is your responsibility to call `free()` on this pointer when you no longer need the memory:

```
void f(int n)
{
    int *array = malloc( enough bytes to hold n integers);
    /* do something with the array */
    free(array);
}
```

Some remarks: First, 'enough bytes to hold n integers' needs some explanation. You may happen to know that on your system, one `int` is 32 bits and thus takes 4 bytes, so you could have written `malloc(4*n)`. However, the size of an `int` is not *quaranteed* to be 32 bits. To find out in a portable way how much memory one variable of a certain type needs, you can use the `sizeof` operator. The following is common C idiom:

```
    int *array = malloc(n*sizeof(int));
```

Although `sizeof` really is an operator (*i.e.*, the brackets around `int` are not necessary), you most often see it used as if it were a function. Note that `malloc` returns a `void *`, which we assign without any casts to an `int *`. This is allowed because C makes an exception for `void` pointers: they can be implicitly converted to pointers to any other type.[2]

Second, you may wonder why we are talking about pointers in the first place, instead of about arrays, and why the example doesn't read

```
int array[] = malloc(n*sizeof(int));
```

This, unfortunately, is invalid in C. I say 'unfortunately' because this adds to the confusion surrounding arrays and pointers, which leads many people to think that arrays and pointers are 'really the same thing' in C. They most definitely aren't (like we saw in the previous section), but they do sometimes appear to be.

For instance, the 'bracket notation' is only 'syntactic sugar' for dereferencing a pointer-with-an-offset. So `a[n]` is actually the same thing as `*(a + n)`. And '`a + n`' when `a` is a pointer and `n` is an integer actually means 'the address `n` items (not bytes!) counting from `a`. This is called 'pointer arithmetic' – you don't have to multiply this `n` by `sizeof(*a)` or anything like that. Note that this is one reason why array indices start at zero.

Since a 'pointer-to-a-variable-of-type-T' is indistinguishable from a 'pointer-to-*several*-adjacent-variables-of-type-T' just by looking at the pointer itself, and because the latter is pretty much how an array is defined, the confusion is quite understandable. Also, an array of only one element is an array too, and an array-of-one-variable looks suspiciously like a pointer-to-one-variable. Especially since it can be addressed both with `*p` and `p[0]`.

The most common source of pointer-array-confusion is the fact that when passing an array as an argument to a function call, the array is not passed in its entirety, but rather a pointer to its first element. In fact, the `average()` function on page 96 might just as well have been defined as

```
double average(double *v, int num)
{
    ...
}
```

This means that when you modify elements of an array which you received as a parameter inside a function, you are modifying the elements of the *caller's* version of the array. This is why 'out value' parameters work as an alternative for returning arrays – see page 109. There are other important implications to this, which will be covered in the next section.

Another very important thing to remember when you are allocating memory yourself, is to call `free()` when you no longer need the memory. Forgetting this leads to what is called a *memory leak*. Any operating system worth its salt will free any memory a

[2]Note that this really is a gaping hole in the type safety of your program. This implicit conversion has been removed in C++, for instance.

program allocated when the program exits, but during the lifetime of your program, it will have to rely on your cooperation. Since memory is a finite resource even on the biggest computers, it will eventually run out if it's never `free()`d. In case `malloc()` cannot allocate the requested memory, it will return zero – or more precisely, the *null pointer*. This special value (usually spelled `NULL`) designates 'a pointer pointing to nothing'. If you try to dereference it, you will get a crash.

For fun, try the program below:

```
#include <stdlib.h>
#include <stdio.h>

int main(void)
{
    unsigned long allocated = 0;
    while (1)
    {
        void *p = malloc(1000);
        if (!p)     /* or: if (p == NULL) */
            break;
        allocated += 1000;
        printf("Allocated so far: %ld bytes\n", allocated);
    }
    printf("Managed to allocate %ld bytes\n", allocated);
    return 0;
}
```

Depending on your computer system, this program may run for quite a while. Also, you may notice something strange: On most platforms, you can allocate more memory than your computer contains! This is because of a clever trick called *virtual memory*[3], in which the operating system swaps out chunks of memory to hard disk when they are not in use, thus freeing up memory, and quickly swaps it back in when your program actually accesses the memory. This is handled transparently by your operating system, and you don't need to modify your program to use it. Note that hard disks are quite a bit slower than RAM, so your program may slow down significantly once it fills up all available physical memory.

It sounds easy to remember: For each `malloc()` there should be a corresponding `free()`. However, memory leaks are quite common in C programs, because sometimes the `malloc()` call is 'hidden' in some function:

```
double *outer3(double a[3], double b[3])
{
    double *array = malloc(3*sizeof(double));
    if (!array)
        return NULL;

    array[0] = a[1]*b[2] - a[2]*b[1];
```

[3] Actually, what is described here is *paging*, which is strictly a separate feature from virtual memory.

```
    array[1] = a[2]*b[0] - a[0]*b[2];
    array[2] = a[0]*b[1] - a[1]*b[0];
    return array;
}
```

(which is why constructs like this are not advisable: they put the burden of `free()`ing all `double` arrays obtained via a call to the `outer3()` function on the 'end user' of that function.) Another cause (and a genuine bug in the code) is that not all code paths reach the `free()`:

```
void f(void)
{
    int *array = malloc(10*sizeof(int));
    /* do something with array */
    if ( some condition)
        return;
    /* do some more with array */
    free(array);
}
```

The latter often happens when a function has 'grown' over time, with exceptions and new code paths being added at a later date; it is easy to forget that you were supposed to call `free()`. For a run-of-the-mill scientific calculation program this is not too much of a problem, but for an application which needs to run for long periods of time (say, a web server) such memory leaks can cause major headaches. Especially since nobody bothers to check the pointer received from `malloc()` for `NULL`-ness (since we have plenty of RAM, so an out-of-memory condition will never happen, right?) and continue to work with the pointer, happily trying to write data through the `NULL` pointer and thus causing a crash.

Another interesting type of bug occurs when you `free()` a pointer but then continue to use it. This may even go unnoticed for quite a while because the data is still there. Most operating systems 'recycle' the memory and may return the same pointer (or a pointer pointing somewhere into the same chunk) at a later `malloc()` call. To catch this kind of bugs, some platforms offer a special 'debug mode' in which a chunk of memory is overwritten with a special bit pattern when it is `free()`d.

This memory management is actually an example of a recurring pattern in programming called *resource management*. Memory is a (scarce) resource, but you can imagine that there are others which you have to *acquire* when you need to work with them, and *release* when you're done. For example, suppose that you are writing a program to operate some external device; you probably have to set up a communication channel with it and 'lock' the device for use with your program, and release it again when you no longer need it. Whereas the operating system is pretty good at releasing any memory you forgot to free, it has no idea that you were supposed to call, say, `release_device()` after you did `acquire_device()`. If you forget it, the communication channel may be unavailable even for your own program when you try to run it again; the device will still be waiting for any sign form the previous program run (which will never come). For cases like this, sometimes the only solution is to reset the device (and/or the computer).

Note that C is sometimes criticized for requiring 'manual memory management', and that some modern programming languages take a different approach. However, this often only works with memory and not with other resources. See §5.12 for some more words on this.

5.8 Passing Arrays To Functions

We passed an array to a function in the very first example of this chapter, but there are a few more things to be said about it.

We mentioned the `sizeof` operator on page 110 saying that `sizeof(a)` gives the number of bytes an instance of type `a` takes up in memory (or if `a` is a variable, the number of bytes this particular variable takes up in memory).

If `a` is an array, it reports the number of bytes taken up by the entire array:

```
#include <stdio.h>

int main(void)
{
    double a[10];
    printf("sizeof(a) = %d\n", sizeof(a));
    return 0;
}
```

will print out 80 (on a system where a `double` is 64 bits, that is). This leads to an interesting trick to find out the number of elements in an array: divide the total number of bytes the array takes up in memory by the size of one element:

```
#include <stdio.h>

int main(void)
{
    double a[10];
    printf("Array size is %d\n", sizeof(a)/sizeof(double));
    return 0;
}
```

will print 10 on any system. You can generalize this a bit and make a macro of it (see §3.11):

```
#define ARRAY_SIZE(a) (sizeof(a)/sizeof(a[0]))
```

You may think that the examples at the beginning of this chapter could be rewritten to use one less parameter:

```
double average(double v[])     /* num no longer needed? */
{
    int num = ARRAY_SIZE(v);
    double sum = 0;
    int n;
    for (n = 0; n < num; n++)
        sum += v[n];
    return sum/num;
}
```

Unfortunately, we're in for a nasty surprise: No matter how large the array is we're passing to the `average()` function, `ARRAY_SIZE` will always be zero (at least on a 32-bit system)! What gives?

The explanation can be found in the previous section, and a hint was already given in the very first section of this chapter: *C doesn't* know *the size of the array once it's passed to a function.* This behavior is sometimes described as 'in C, arrays *decay* to pointers when passed as a function parameter'.

So the `average()` function above, to C, looks like (and might as well have been written as)

```
double average(double *v)
{
    int num = ARRAY_SIZE(v);
    ...
}
```

and the `sizeof` a pointer is 4 on a 32-bit system. Therefore, the `ARRAY_SIZE` macro calculates the `sizeof` a pointer divided by the `sizeof` a `double`, which is 4/8, yielding zero because of truncation. Had you passed an array of `int`s, you'd probably have gotten 1 as the result on a 32-bit system, and 2 on a 64-bit system. All quite interesting results, but certainly not what you had expected.

The moral of this section: You cannot infer the size of an array once it has been passed to a function, so you need to pass its size as one extra argument.

5.9 A Useful Example: Factors

In this section, we will build a complete program which does something useful: Given an integer number, print out its factors. If there are none, the number is prime.

To make the program 're-usable', we would like to have it get the number from the user somehow, so we won't have to recompile a separate version of the program for each number we would like to investigate.

Perhaps you remember that in §3.12, it was alluded that there is a different version of the `main()` function which takes parameters. The first step is to construct a program which uses that, and it looks like this:

```
#include <stdio.h>

int main(int argc, char *argv[])
{
    int i;
    for (i = 0; i < argc; i++)
        printf("argv[%d] = %s\n", i, argv[i]);
    return 0;
}
```

The two parameters to main() are the *argument count* (*i.e.* the number of arguments passed to the program) and the *program arguments*. The 'char *argv[]' construct may look a little intimidating, but it simply means 'an array of char *', in other words an array of strings.

Most operating systems, when you start a program, pass the program arguments to the program this way. Usually, the program arguments are separated by white space, and the first (actually, zero-th) argument is the program name itself. So if you type

```
myprogram foo bar
```

then the main() function of 'myprogram' will be called with argc equal to 3, argv[0] equal to "myprogram", argv[1] equal to "foo", and argv[2] equal to "bar" (without the quotes – these are to signify that they are zero-terminated C strings).

Our 'factor' program needs a *number* though, and not a string. The next step is to make a program which checks whether there is exactly one program argument, converts it to an integer, and prints it:

```
#include <stdio.h>
#include <stdlib.h>

int main(int argc, char *argv[])
{
    int number;
    if (argc != 2 || !strcmp(argv[1], "-h"))
    {
        printf("Usage: %s <number>\n", argv[0]);
        printf("Prints out the factors of <number>\n");
        return 1;
    }

    number = atoi(argv[1]);
    printf("We will investigate %d\n", number);
    return 0;
}
```

The first part of the program checks whether the user has actually entered a program argument. It is common practice to print out a short description of how the program is used otherwise. Note how we put argv[0] in there, so it will still print out the program's name even if the user has renamed it. As an extra feature, we've made sure that the program also prints out this information when invoked with a '-h' (for 'help') argument, which is also common (at least for programs on UNIX systems). You could make the program even more friendly by also checking for /? since that's the common 'help' switch used in DOS and Windows. Also, note that we're using the fact that C does 'short-circuiting' (see page 73), so that the argv[1] is only evaluated if there really *is* an argv[1].

The atoi() function is a brother of atof() which we already met before. This one converts its string argument to an integer value.

Now, on to the 'meat' of our program. Instead of printing out the number, we will call a function print_factors(). That function could look something like this:

```
#include <math.h>

int print_factors(int n)
{
    int found = 0;
    int i;
    for (i = 2; i <= sqrt(n); i++)
    {
        if (n % i == 0)
        {
            found++;
            printf("%d = %d x %d\n", n, i, n/i);
        }
    }
    return found;
}
```

Note that we use the modulo operator (see page 49) to determine whether i divides n, that we start at 2 (1 is always a divisor), that we #include <math.h> and use sqrt(n) as an upper limit of divisors to try (remember that one of the factors p, q is less than $\sqrt{n}$ if $p \times q = n$ unless p and q are *both* equal to $\sqrt{n}$), and that we keep track of the number of factors we found. That way, if print_factors() returns zero, the number was prime. We can use that in our main() function:

```
    if (print_factors(number) == 0)
        printf("%d is prime!\n", number);
```

5.10 Variable Length Arrays

It was mentioned in §5.7 that variable length arrays (VLAs) were added to the C99 standard. Not all compilers support this (yet) but this section gives an overview of this new feature.

In C99, a function such as the following is legal:

```
void vlafunc(int n)
{
    int vla[n];     // illegal in C89, OK in C99
    ...
}
```

There are some restrictions to VLAs (if you must know: they cannot be `extern` or `static` – things we haven't seen yet and which will be covered in chapter 10, and `n` must (obviously) be a positive integer or a (runtime) error will occur).

The prototype of a function *taking* a VLA as one of its parameters uses an asterisk:

```
double avg_of_vla(int size, double v[*]);
```

but in the definition of it, the syntax is slightly different:

```
double avg_of_vla(int size, double vla[size])
{
    double sum = 0;
    for (int i = 0; i < size; i++)
        sum += vla[i];
    return sum/size;
}
```

Multi-dimensional VLAs are allowed too:

```
void vlafunc2(int n, int m)
{
    int vla2d[n][m];
    ...
}
```

Also, note that the `ARRAY_SIZE` problem when looking at an array passed to a function (see §5.8) unfortunately is not solved by VLAs (although it does report the correct array size in the scope the VLA was declared in), so you still need to pass the size of the array as a separate parameter.

5.11 Synopsis

A *pointer* can be described in 'high level terms' as a *reference* to a variable (rather than the variable itself) or in 'low level terms' as the *memory address* of a variable (rather than the *contents* of that address). Pointers, in C, are designated by the `*` symbol, and you can obtain a pointer to an existing variable by prepending the 'address-of' operator, `&`.

Pointers in C are *typed*: a pointer-to-an-`int` is different from a pointer-to-a-`float`. Thus, pointers are more than just memory addresses. There is one exception: a `void*`, which is a pointer-to-*something* (rather than a pointer-to-*nothing* like the name suggests).

Dereferencing a pointer uses the * prefix operator: If `p` is a pointer-to-an-`int`, then `*p` is the value of that `int`.

When doing *pointer arithmetic* (adding an integer to a pointer), the unit of calculation is the size of the pointee. That means that it is easy to handle multiple variables of the same type stored in contiguous memory: If `p` is a pointer to the first of such a set of variables, then `p + n` is a pointer to the `n`th variable, and `*(p + n)` is the value of the `n`th variable. Such a set of variables is called an *array*, and a convenient shorthand notation for `*(p + n)` is `p[n]`. This *bracket notation* is also used when declaring an array: `double a[10];` declares an array of ten `double`s (numbered `a[0]` through `a[9]`).

You can declare multi-dimensional arrays by appending dimensions in brackets: `double A[2][3];` declares a two-by-three matrix, and A_{ij} is addressable as `A[i][j]`.

In C, a *string* is understood to be a zero-terminated array of `char`s.

When an array is passed to a function, it *decays* to a pointer, meaning you cannot infer its size anymore. The `sizeof`-operator (which yields the number of bytes its operand takes up in memory) simply returns the size of a pointer on the current system in that case.

You can *allocate* a chunk of memory using the `malloc()` function, which returns a (`void`) pointer to it. You can use the `sizeof` operator to find out how many bytes are needed for one instance of a certain type. It is your responsibility to `free()` the memory when it is no longer needed.

When you define your `main()` function as `main(int argc, char *argv[])`, the operating system will fill in the `argv` array-of-strings with the program arguments as given on the command line when invoking your program. The number of elements in this array is given by `argc`. On most operating systems, `argv[0]` is set to the name of the executable itself.

5.12 Other Languages

'Manual' memory management is considered 'old-fashioned' in computer science, and many modern languages do away with it. Some languages (like Java and C#) use a *garbage collector*: a runtime system keeps track of which pointers are still in use, and automatically schedules chunks of memory to be freed when there are no references to them anymore. Other languages (like C++) offer *deterministic destructors* – special pieces of code which get called as soon as a variable 'falls out of scope'; if this variable represents a chunk of memory the destructor (which is guaranteed to be called) can free it. The nice thing about this is that it also works for other resources than just memory (such as the example on page 113: the `release_device()` would be called in the destructor), and when applied carefully you will never have to worry about releasing resources.

The concept of zero-terminated strings is very much a 'C-ism'. Another way of handling the size of a string is by prepending it with its length (usually in a way which is 'hidden' to the programmer). For instance, Pascal does this (in some 'dialects' of

Pascal, there was one single byte prepended to the string, which limited the maximum string length to 255 characters). Many languages offer higher-level string handling by making a string a first-rate datatype and offering special operators for concatenating and splitting strings, etc. In those languages, strings *can* often be compared using 'normal' == operators.

In C#, multi-dimensional arrays are indexed in a subtly different way than in C and C++: `a[i][j]` is written as `a[i,j]` there. This can be confusing when switching back and forth between these languages, since as mentioned on page 103, `a[i,j]` is a valid expression in C/C++ as well (yet with a different meaning).

In Pascal, the size of an array is part of its type; this has the advantage that you can never accidentally pass an array-of-three-items to a function expecting an array-of-four-items, but it has the drawback that you can never make a generic function operating on arrays of any size.

Some languages (notably BASIC and FORTRAN start their array indices with 1 instead of 0, and some languages allow arbitrary indices, with a syntax like `array(3..7)` which denotes an array which has elements 3 through 7.

Some languages or compilers have built-in bounds checking; in some cases this can be switched off for the final 'release build' of the software (after ample testing, of course) to remove the performance penalty.

5.13 Questions and Exercises

5.1 Write a `strlen()` implementation yourself. It should return the number of characters in the string passed to it, not counting the terminating zero. To prevent your compiler from getting confused with the real `strlen()`, call yours `mystrlen()` or something to that effect.

5.2 Write an `atoi()` implementation yourself.

5.3 Write a function to count the number of words in a string. It should be robust against multiple spaces in a row (*i.e.*, it should not simply count spaces).

5.4 Use the `fac()` function you wrote in exercise 4.3 to make a `fac` *program* which reads a number via `argv` and prints out its factorial on the screen, similar to the `factor` example in §5.9.

5.5 ⋆ Write a function which sorts an array of numbers in ascending order.

Chapter 6
Data Types and Structures

The structure of language determines not only thought, but reality itself.

Noam Chomsky

6.1 Levels of Abstraction

With the data types available in C, there is quite a bit you can do. However, these data types are very much a reflection of what the actual hardware offers, and not so much what a scientist would have in mind. After all, a scientist would likely think in terms of 'ordinals' and 'real numbers', and not so much in terms of 'a number represented in floating point with extended precision'. In other words, C offers an abstraction layer 'from the computer upwards': you can give variables sensible names, and can write expressions (almost) like you would in ordinary math, but you are still bound to what the computer hardware has to offer.

This is much in the spirit of C. For example, common CPUs don't have 'strings' as a native data type, so C doesn't pretend they do. On the other hand, this can be seen as a (major) drawback of C, since it doesn't 'free' the programmer from thinking about the *computer* instead of the *problem*.

C does allow you to extend its type system programmatically by defining your own data types. These 'custom types' can help improve the expressive power of your programs tremendously, since they are treated by the compiler (almost) like they were built-in types. This chapter focuses on these features of C.

6.2 Type Definitions

The first (and most straight-forward) way of defining a new type in C is to give a new name to an existing one. This may not seem very useful (after all, there isn't much 'new' to a type defined this way) but it does make a (small) step towards being able to stay 'in the problem domain' while writing your program.

C offers the typedef keyword for this purpose, and it is used with the following syntax:

```
typedef old new;
```

in which *old* is an 'existing' type, and *new* its new name (also called its 'alias').

Suppose you have a function which returns the 'size' of something. By definition, a size cannot be negative, so you would use an unsigned type. A good example is the strlen() function, which we have seen before. If you look in your system header files, you may find out that the prototype of strlen() actually looks like this:

```
size_t strlen(const char *str);
```

(The const tells you that this function will not modify the string you pass it as a parameter – see page 128.) The return type of the function, size_t, is not a native C type. Instead, it is defined like so:

```
typedef unsigned long size_t;
```

It is good practice, when defining new data types, to append '_t' when you are using a very common name (such as 'size').

The fact that the new type defined this way is just a new name for an existing type is both an advantage and a drawback. The advantage is that you can use the 'new' type in ordinary arithmetic expressions, and that you can call a function which was defined as taking the 'old' type as a parameter with your 'new' type (and the other way around, too). This is also a drawback, because this still doesn't prevent you from making semantic mistakes (or in more blunt language: introducing bugs in your code). Suppose you do this:

```
typedef double mass_t;
typedef double velocity_t;
typedef double distance_t;
typedef double energy_t;

energy_t kinetic(mass_t m, velocity_t v)
{
    return m*v*v/2;
}
```

in an attempt to prevent accidentally calling the function with 'mass' and 'velocity' interchanged, you're in for a disappointment. As far as the compiler is concerned, you can pass any double as argument to the kinetic() function:

```
mass_t m = 1.5;
distance_t d = 3.14;
energy_t e = kinetic(d, m);
```

There *are* ways to enforce this 'strong typing' in C, and we'll look at one in the next section. However, the result is clumsy and impractical.

Also, typedefs do not provide a solution to the 'array-of-incorrect-dimension' problem. You are of course free to do the following to increase the readability of your code:

```
typedef double vector3[3];

double inner(vector3 a, vector3 b)
{
    return a[0]*b[0] + a[1]*b[1] + a[2]*b[2];
}

int main(void)
{
    vector3 a = {1, 2, 3};
    vector3 b = {4, 5, 6};
    vector3 c = inner(a, b);
    printf("(%f, %f, %f)\n", c[0], c[1], c[2]);
    return 0;
}
```

but you can still call your inner() function with the wrong type of array:

```
double d[2] = {1, 2};
vector3 e = inner(a, d);     /* boom! */
```

The solution presented in the next section for this particular issue is not-so-clumsy, and quite practical. It also solves the problem that you cannot return an array from a function, not even if it is 'masqueraded' with a typedef:

```
vector3 outer(vector3 a, vector3 b)
{
    vector3 result;
    result[0] = a[1]*b[2] - a[2]*b[1];
    result[1] = a[2]*b[0] - a[0]*b[2];
    result[2] = a[0]*b[1] - a[1]*b[0];
    return result;
}
```

This will still not compile, and any tricks you may come up with involving returning a double* instead of a vector3 suffer from the same problems as mentioned in the previous chapter.

6.3 Data Structures

A very powerful feature of C is the possibility of defining 'collections of data'. These are called structs. Unlike arrays, the members in a struct can be *heterogeneous*, *i.e.* they can be of different types, and they are *named* (instead of just 'numbered'). As an example of a *homogeneous* struct, let's redefine the three-dimensional vector:

```
struct vector3
{
    double x;
    double y;
    double z;
};
```

We can now define our `inner3()` function like this:

```
double inner3(struct vector3 a, struct vector3 b)
{
    return a.x*b.x + a.y*b.y + a.z*b.z;
}
```

You see that you can use the 'dot-operator' (`.`) to get at the named members of the `struct`.

Also, whereas it gave all kinds of trouble trying to return an array from a function, returning a `struct` is just fine:

```
struct vector3 outer3(struct vector3 a, struct vector3 b)
{
    struct vector3 result;
    result.x = a.y*b.z - a.z*b.y;
    result.y = a.z*b.x - a.x*b.z;
    result.z = a.x*b.y - a.y*b.z;
    return result;
}
```

The only chore is having to type '`struct`' all the time, and this can be taken care of by using a `typedef`:

```
typedef struct
{
    double x;
    double y;
    double z;
} vector3;
```

after which the rest of our code would look like this:

```
double inner(vector3 a, vector3 b)
{
    return a.x*b.x + a.y*b.y + a.z*b.z;
}

vector3 outer(vector3 a, vector3 b)
{
    vector3 result;
    result.x = a.y*b.z - a.z*b.y;
```

```
    result.y = a.z*b.x - a.x*b.z;
    result.z = a.x*b.y - a.y*b.z;
    return result;
}

int main(void)
{
    vector3 a = {1, 2, 3};
    vector3 b = {3, 1, 4);
    vector3 c = outer(a, b);
    printf("(%f, %f, %f)\n", c.x, c.y, c.z);
    return 0;
}
```

Note that a `struct` can be initialized with an *initializer list* the same way an array can (using the {···} notation). Also, note that we got rid of the 3's in the `inner()` and `outer()` function names, because there is no longer a need to 'remind' us of the dimension: Given a two-dimensional vector

```
typedef struct
{
    double x;
    double y;
} vector2;
```

the following code will no longer compile:

```
vector3 a = {1, 2, 3};
vector2 d = {2, 7};
vector3 c = outer(a, d);
```

It will say something like '`incompatible type for argument 2 of 'outer''` – which is a good thing!

If you specify fewer elements in the initializer list than there are members in the `struct`, the remaining members are initialized to zero.

In C99, you can initialize specific members of a `struct` by naming them. For example, you could initialize only the `z` and `y` members of a `vector3` variable like this:

```
vector3 e = { .z = 4, .y = 2 };
```

6.3.1 'Fake' Strong Typing

It was promised in the previous section to take a look at increasing the type-safety of your programs. With C, you're stuck between a rock and a hard place when it comes to using user-defined types for this purpose.

On the one hand, you can define simple aliases for existing types to increase readability, but when two types (say, `mass_t` and `velocity_t`) are both `typedef`'ed aliases for `double`, then you can still accidentally mix them up (and in fact, *any* `double` will do).

On the other hand, we have `structs` which the compiler does check very strictly, but which are rather clumsy to use for this purpose. Suppose we do something like

```
typedef struct
{
    double value;
} mass_t;

typedef struct
{
    double value;
} velocity_t;

typedef struct
{
    double value;
} energy_t;

energy_t kinetic(mass_t m, velocity_t v)
{
    ...
}
```

then you *can't* call the `kinetic()` function with mass and velocity in the wrong order (even though the structures are all exactly the same). The drawback is that using these types is clumsy: You can no longer type

```
energy_t kinetic(mass_t m, velocity_t v)
{
    return m*v*v/2;
}
```

because the compiler doesn't know what it means to multiply these data structures. Unlike in other, more modern programming languages (most notably C++), you cannot define your own operators in C. Instead, you'll have to say:

```
energy_t kinetic(mass_t m, velocity_t v)
{
    energy_t e;
    e.value = m.value*v.value*v.value/2;
    return e;
}
```

Also, the syntax for using these types is a little strange, because you'll have to write

```
mass_t m = { 42 };
```

instead of just

```
mass_t m = 42;
```

and you cannot call your `kinetic()` function with literal values anymore.

These drawbacks make that this 'trick' of forcing stronger type-safety in C programs is hardly ever used.

6.3.2 Call-by-Name vs. Call-by-Value

It was mentioned earlier that when you pass a variable as a parameter to a function, the function receives a *copy* of that variable. If it changes the value of this variable, the 'original' (on the caller's side) remains unchanged. This is sometimes called 'call-by-value'.

The exception was when passing an *array*, because when passing an array as a function argument, it 'decays' into a pointer. Calling functions with pointers to parameters instead of copies of those parameters is sometimes called 'call-by-name', because it is as if you pass the *name* of a variable to the function; the function code then 'looks up' this variable in memory and can find its value there, but may also modify it.

It may appear that passing copies around is always better (with the exception of specific functions which are *supposed* to modify their parameters), but there is another reason to prefer the other mechanism of passing parameters to functions, namely performance.

Take the `vector3` example as mentioned above. To call the function

```
double inner(vector3 a, vector3 b)
{
    return a.x*b.x + a.y*b.y + a.z*b.z;
}
```

with two `vector3`s, the resulting program will (under the hood) create copies of the two vectors you're calling the function with, place them in a special area of computer memory (called the *stack*), and call the function. This may seem innocuous, but on most computer systems, one `double` takes at least 8 bytes, so to pass those two vectors, 48 bytes need to be transferred.

Now contrast this with passing *pointers to vectors* instead:

```
double inner(vector3 *a, vector3 *b)
{
    return a->x*b->x + a->y*b->y + a->z*b->z;
}
```

in this case, only 8 bytes need to be transferred (on a 32-bit machine, where pointers take 4 bytes each). Note the different syntax: `a->x` is shorthand for `(*a).x`, *i.e.* the 'arrow' is used when you have a *pointer* to a struct where you would have used the 'dot' if you had the struct *itself.*

Now, the problem is that if you only look at the prototype of this new, pointer-based function, you cannot be sure that it will not modify its parameters. For this, C has the const modifier, with which you can 'promise' that you will not modify whatever is being pointed to:

```
double inner(const vector3 *a, const vector3 *b);
```

Unfortunately, it is 'implementation-defined' what happens when a program modifies such a const variable anyway. Usually, you'll at least get a compiler warning (but that's not guaranteed). C++ made it explicitly illegal, and a C++ compiler will treat it as an error when a function tries to modify a const parameter.

Incidentally, we have met const before: It was used in the prototype for the printf() function (see page 56) and the strlen() function (see page 122) to notify the caller that the strings passed as arguments wouldn't be modified by the functions.

6.3.3 Heterogeneous Structures

Apart from being *named* instead of *numbered* (as with arrays), C structs need not contain members of the same type:

```
typedef struct
{
    double mass;
    double distance_from_sun;
    char name[MAX_PLANET_NAME];
} planet_t;
```

This is also a nice example of the problem with strings in structs: Either you pick a fixed length, with the risk of getting in trouble when a newly discovered planet is given a longer name[1], or you change the struct like so:

```
typedef struct
{
    double mass;
    double distance_from_sun;
    char *name;
} planet_t;
```

In this case, you will need to allocate room for the name when you fill in a new planet_t, and you will also need to remember to free() the name field again when the struct is no longer needed. For complex structures like these, you would usually provide a constructor/destructor function pair:

```
planet_t *create_planet(double m, double d, const char *name)
{
    planet_t *p = malloc(sizeof(planet_t));
    p->mass = m;
    p->distance_from_sun = d;
    p->name = strdup(name);
```

[1] A reasonably small 'risk' in this case, but you get the point

```
    return p;
}

void destroy_planet(planet_t *p)
{
    free(p->name);
    free(p);
}
```

The `strdup()` means 'duplicate string' and is basically a shortcut for

```
char *strdup(const char *s)
{
    char *new_string = malloc(strlen(s) + 1);
    if (new_string != 0)
        strcpy(new_string, s);
    return new_string;
}
```

Given a definition for `planet_t` and this constructor/destructor pair, you would never create planets 'manually'[2], but would always use `create_planet()` to create one, and use `destroy_planet()` when you no longer need it. This way, when you later decide to extend the `planet_t` structure (say, with another string), you will only need to change the constructor/destructor functions and not have to modify your code all over the place.

By the way, the choice to have the constructor return a *pointer* to a `planet_t` and the destructor take a *pointer* instead of a `planet_t` is rather arbitrary. The advantage is that this prevents one copy from being made (see the previous subsection). The disadvantage is that you may be tempted to call `destroy_planet()` on a stack-based variable (see page 84), like so:

```
{
    planet_t p;
    p.mass = 5.976e24;
    p.distance_from_sun = 149.6e9;
    p.name = strdup("Earth");
    // some more code
    destroy_planet(&p);
}
```

This leads to undefined behavior, because you `free()` something which was never `malloc()`ed. It would have been a similar problem if you forgot the `strdup()` but simply wrote

```
    p.name = "Earth";
```

because in that case too, `destroy_planet()` would try to `free()` the `name` string – which was never `malloc()`ed either.

[2]Unless your name is Slartibartfast, of course

It is a common convention that a function starting with 'create' and returning a pointer allocates the memory for the thing it creates. In that same convention, it is best for the 'destroy' function to take care of free()ing not only the members of the structure it is passed, but also the structure itself. This is to keep symmetry in malloc() and free() calls in the client code.

If a function doesn't *create* a structure but simply *fills* it with relevant data, convention is that this function starts with something like initialize or init. A function which would take care of cleaning up all the *members* of a structure but not the structure *itself* would be called something with deinitialize (despite the clumsy and casuistic flavor of that word). The advantage of this convention that it is immediately apparent to someone using your functions whether he is allowed to pass stack-based structures, and whether he is supposed to manually free() the structure afterwards.

6.3.4 Complex Data Structures

You can define structures which contain other structures. Suppose you are writing a particle interaction simulation; in that case, a particle could be defined by its mass, its position in space, and its current velocity:

```
typedef struct
{
    double mass;
    vector3 position;
    vector3 velocity;
} particle_t;
```

To get at the x-position of a certain particle p, you would write p.position.x (*i.e.* after the first dot you select the 'position' field of the particle p, and after the second dot the x field of this vector).

In fact, you can define 'self-referential' structures, which contain a pointer to a structure of the same type. There are several applications for this, and we'll examine one here.

The remainder of this subsection pertains a (lengthy) example of a name-address data base type of application[3]. Suppose you have defined a structure for each data entry:

```
typedef struct
{
    char *name;
    char *address;
} person_t;
```

along with an initialize/deinitialize pair:

[3]This type of example is probably not something a 'scientifically inclined' person would be naturally interested in, but bear with me: it does illustrate some points very well.

```
void init_person(person_t *p, const char *name, const char *address)
{
    p->name = strdup(name);
    p->address = strdup(address);
}

void deinit_person(person_t *p)
{
    free(p->name);
    free(p->address);
}
```

We could now set up a simple database application. For the sake of simplicity, forget about storing the data on disk, and assume we're writing an application which will simply always be running. In our first attempt, the crux of our 'database' would be a simple array of `person_t`, which we'll put in the global scope (see §4.7, page 85), but we're immediately faced with a similar problem as when we had to determine a 'maximum name length':

```
#define MAX_NUMBER_OF_PEOPLE 1000
person_t database[MAX_NUMBER_OF_PEOPLE];
int current_index = 0;
```

Let's stick with this fixed number for now; we'll revisit this later.

Inserting new people in the database would be rather straightforward (we'll take care of the exceptional situation that the database is full by returning a special error code, which you'd actually put in the *header* file of your database code and not in the *implementation* file, so that 'client' code (*i.e.*, code which *uses* the functions we're defining here) has access to it):

```
#define ERROR_DATABASE_FULL (-1)

int insert_person(person_t p)
{
    if (current_index == MAX_NUMBER_OF_PEOPLE)
        return ERROR_DATABASE_FULL;
    database[current_index++] = p;
    return 0;
}
```

Here, the `person_t` structure is copied into the database array. Note that this happens via a simple assignment, as if we were talking about `int`s or `double`s – in such an assignment, all the members of the `struct` are copied over. Also note the '++' in `current_index++`. We have seen the meaning of this operator before (see page 77), but here it is used *inside* the brackets of the `database` array. It means that the `current_index` is incremented *after* `p` is assigned to the current item in `database`, and it is equivalent to

```
    database[current_index] = p;
    current_index++;
```

This is called *post-increment*, and there's also *pre-increment:* When the ++ operator is prepended to the variable, the value is incremented *before* it is used. So

```
    database[++current_index] = p;
```

would be equivalent to

```
    current_index++;
    database[current_index] = p;
```

which, in this case, is not what we want (because index 0 would be skipped). Post- and pre-increment is idiomatic to C programmers, and it is mentioned here because it is very often used when filling or iterating over arrays.

Also, note that this function can be made more efficient by passing a (`const`) *pointer* to a new `person_t` and dereferencing it inside the function:

```
int insert_person(const person_t *p)
{
    if (current_index == MAX_NUMBER_OF_PEOPLE)
        return ERROR_DATABASE_FULL;
    database[current_index++] = *p;
    return 0;
}
```

because in our original version, *two* copies are made each time a new person is inserted into the database: one to do the actual insertion, and one extra time just to call the function itself.

It is important to realize that 'copying a `person_t`' means that only the *direct* contents of this `person_t` structure are copied (this is called a *bitwise copy*, because the actual bit pattern making up the contents of the `struct` is copied over). In other words, the copy will contain pointers to the *original* `name` and `address` strings. This is very important, because when you call `deinit_person()` on one of these `person_t` structures, it will `free()` those strings, leaving the other `person_t` with so-called 'stale pointers', *i.e.*, pointers to data which is no longer valid. The following picture illustrates the 'two pointers to the same data' situation:

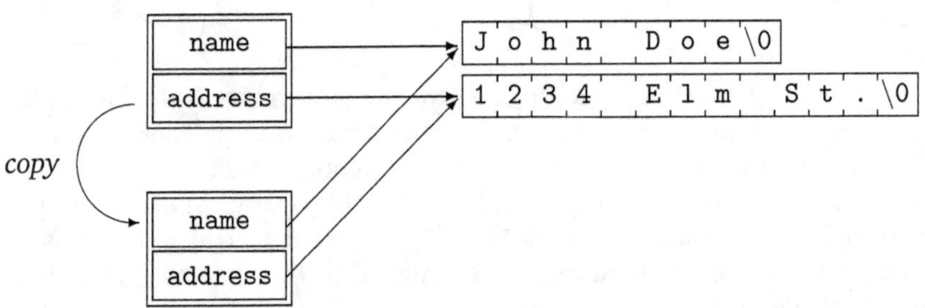

A function which would copy all referenced data as well is said to perform a *deep copy*, and in that context a bitwise copy is also sometimes called a *shallow copy*.

Finally, a function needs to be defined to retrieve a person from the database, most likely based on that persons name:

```
const person_t *find_person(const char *name)
{
    int i;
    for (i = 0; i < current_index; i++)
        if (!strcmp(database[i].name, name)
            return &database[i];
    return NULL;
}
```

What this function does is simply iterate over all persons in the database, compare their `name` field with the given name, and return a pointer to the correct `person_t` in case they match (and `NULL` if no matching person could be found – see page 112). The actual implementation of a real application using these functions is left as an exercise, and we will instead focus on a few problems with this initial approach.

The most obvious one is the problem with `MAX_NUMBER_OF_PEOPLE`. If you pick a large number, you will waste a lot of memory when your database is used for a modest amount of people (although this is less of a problem with a dedicated application). If you pick a number too small, you'll run the risk of reaching the limit too soon.

One solution to this problem is to use a *dynamic* array, using the techniques explained in section 5.7. Instead of a fixed-size array, we put a pointer in the global scope, and make sure it points to an 'adequately-sized' chunk of memory. Client code doesn't need to be changed (which is good):

```
person_t *database = NULL;
int current_index = 0;
size_t current_size = 0;
```

This simply means that `database` is a pointer to one or more `person_t` entries, and it is initialized to `NULL` (*i.e.*, there is no memory allocated for the database yet).

The function used to insert new people in the database needs to do a few things differently from our initial version:

1. When called for the first time, it should initialize the database and allocate room for an initial amount of people (to be defined), and update `current_size` to reflect this number.

2. When called subsequently, check whether the current database is full (*i.e.*, whether `current_index` has reached `current_size`), and if so, allocate room for more.

3. Insert the new person at the right place in the database.

The following function fulfills these requirements:

```
#define CHUNK_SIZE 100
#define ERROR_OUT_OF_MEMORY (-2)

int insert_person(const person_t *p)
{
    if (!database)      /* first call - initialize database */
    {
        database = malloc(CHUNK_SIZE*sizeof(person_t));
        if (!database)
            return ERROR_OUT_OF_MEMORY;
        current_size = CHUNK_SIZE;
    }

    if (current_index == current_size)      /* database full */
    {
        size_t new_size = current_size + CHUNK_SIZE;
        person_t *new_space = realloc(database,
                                      new_size*sizeof(person_t));
        if (!new_space)
            return ERROR_OUT_OF_MEMORY;
        database = new_space;
    }

    database[current_index++] = *p;
    return 0;
}
```

We see a new function here, `realloc()`, which is from the `malloc()` and `free()` family. It tries to change the size of the allocated chunk of memory; if that succeeds, the result is a pointer to this new memory. If it fails, it will return `NULL` (which the function above translates into an appropriate error code, the definition of which would be put in the header file again so client code can check for it); in that case, the original memory will be left unchanged.

Basically, `realloc(p, s)` is functionally equivalent with a `malloc(s)`, followed by a `memcpy()` ('memory copy') of the original data to the new space, followed by a `free(p)` to clean up the original space. However, `realloc()` may be able to do this more efficiently (without the extra copy) because it 'knows' about the internals of the memory allocation scheme, and if the requested extra room happens to be available adjacent to the original allocated memory, it could simply 'extend' the original chunk.

You can see that when the database is full, we use this `realloc()` function to make new room. In the implementation above, we extend it with a fixed amount each time (namely, `CHUNK_SIZE`). An alternative would be to extend the memory not by a fixed *amount*, but by a fixed *factor*. Investigation of many existing real-world systems has shown that a good value for such a factor is 1.5.

Whichever approach is taken for the dynamic resizing of the database, note that the `find_person()` function does not need to change, because locating a `person_t` at a specific index is the same for arrays and dynamically allocated memory. Even if someone calls `find_person()` before the first person is added to the database, the function will work 'as advertized', because the `for`-loop would immediately terminate.

This more dynamic implementation solves the problem of deciding a good size for the database, but it leaves a few other, more interesting problems (or at least, problems more pertaining to the subject of this subsection). When the database grows, finding people in it gets slower and slower, because the algorithm for finding people is *linear:* It simply iterates over *all* `person_ts` in the database until one is found which matches. In the notation explained in §4.8.2, this is an $\mathcal{O}(N)$ algorithm, and that's not very good.

A solution to this is to change the way the database is organized from a simple unordered array into something which facilitates searching, for instance by making sure the list of people is alphabetically sorted. In that case, we can use the bisection algorithm (also called a *binary search*) explained in §4.8.2, which is $\mathcal{O}(\log N)$.

Another problem, which we'll discuss first, is that of removing entries from the database. When using an array as the data storage for the database, there are two possible ways to remove entries: One is to *really* remove them, by locating the entry to be removed and copying the rest of the items one slot 'to the left':

```
#define ERROR_PERSON_NOT_FOUND (-3)

int remove_person(const char *name)
{
    int i;
    for (i = 0; i < current_index; i++)
        if (!strcmp(database[i].name, name)
            break;

    if (i == current_index)
        return ERROR_PERSON_NOT_FOUND;

    deinit_person(&database[i]);
    for (; i < current_index - 1; i++)
        database[i] = database[i + 1];

    current_index--;
    return 0;
}
```

We'll first look at a few idiomatic C constructs which are used in this function. Note the use of the `break` statement, which we've seen in §4.2 in conjunction with the `switch` statement, but which can also be used inside a loop to literally 'break out'. If the correct person is found, we use it to terminate the loop immediately. The next piece of code compares the value of the index `i` with the 'normal' end criterion of the `for`-loop, to see whether we looped over the entire database (in which case the person with the given name apparently wasn't in the database, so we'll return this as an error code).

Next, the entry at the correct index is cleaned up using the `deinit_person()` function, which frees all memory taken up by the entry (but not including the memory taken up by the `struct person_t` itself).

After that, we have to 'shrink' the database, and we do that by setting up a new `for`-loop starting at the current value of `i` (note that the first clause of the `for` statement is empty, so the loop is entered with the value of `i` unchanged) and looping over the rest of the database, each time replacing the current entry with the next. Also, note that this loop continues until `i` equals `current_index - 1`. Alternatively, we could have placed the `current_index--` statement *before* the loop, but it is better to be explicit about it in the `for`-loop itself, because the loop is referring to items at index $i + 1$. The figure below illustrates the procedure. Each box stands for one entry in the database array ('ci' stands for `current_index`).

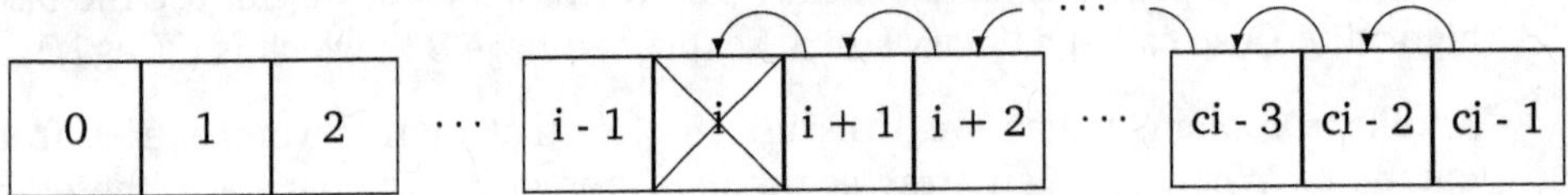

This method for 'shrinking' the database is relatively expensive: If the person to be removed is near the end of the database, it'll take many comparisons to find the corresponding entry; if that entry is near the beginning of the database, it's found relatively quickly but it'll take a lot of copying to move all the entries after it by one slot to reclaim its space. Therefore, another approach may be beneficial.

Instead of *really* removing entries from the database, it is enough to only *mark* them as no longer valid. In that case the other functions need to be modified as well (to skip over such entries). Let's look at the `remove_person()` function first:

```
int remove_person(const char *name)
{
    int i;

    for (i = 0; i < current_index; i++)
        if (!strcmp(database[i].name, name))
            break;

    if (i == current_index)
        return ERROR_PERSON_NOT_FOUND;

    deinit_person(&database[i]);
    database[i].name = NULL;
}
```

To mark an entry as removed, we simply put a `NULL` in the `name` slot of the database array. Of course, the `find_person()` function needs to be made aware of this:

```
const person_t *find_person(const char *name)
{
    int i;
```

```
    for (i = 0; i < current_index; i++)
        if (database[i].name && !strcmp(database[i]->name, name))
            return database[i];

    return NULL;
}
```

(Remember from page 73 that the `strcmp()` will only be called if the first condition in the `if`-statement evaluates to non-zero, because otherwise the `if`-statement would be 'short-circuited'.)

Also, the `insert_person()` function should be modified; instead of simply appending the new person to the database, it should first iterate over the entire database looking for entries marked as removed, and re-use that slot in the database. Otherwise, many insertions and removals will make the database grow indefinitely. Unfortunately, this means that the `insert_person()` function becomes quite a bit more expensive computationally!

Luckily, there are alternative data structures with different characteristics. The next section gives some examples.

6.3.5 Alternatives to Arrays

As pointed out in the previous subsection, using a simple array containing the `structs` of the database has some drawbacks, mainly in the performance department. This subsection will offer a solution which is convenient in a lot of similar problem areas, by separating *what* is stored from *how* it is stored. The simple array may be the first thing to come to mind when designing a data structure to hold multiple entries, but it need not be the best choice.

There has been much research in computer science on data structures, and there are several 'patterns' which have emerged as being very useful. A very important one is the so-called *linked list*. In such a list, each entry (in this context also called a *node*) contains not only its data, but also a pointer to the next entry. In the case of our `person_t`, this would look somewhat like this:

```
typedef struct person
{
    char *name;
    char *address;
    struct person *previous;
} person_t;
```

Note that this structure contains a pointer to an entry *of the same type*. This is perfectly legal, but do note that we couldn't use the name `person_t` inside the structure definition (after all, it wouldn't have been defined at that point yet), so we have to give the `struct` a name (also sometimes called a 'tag'); in this example it's '`person`'.

This 'self-referential' definition is legal because it only contains a *pointer* to the same `struct`, not the `struct` itself. The latter would be illegal, because it would lead to an infinite recursion.

This new type of person_t is not supposed to be placed in an array, but to 'live on its own' in memory. To reflect that, we will use a create/destroy pair instead of the init/deinit pair we had before:

```
person_t *create_person(const char *name, const char *address)
{
    person_t *new_person = malloc(sizeof(person_t));
    if (!new_person)
        return NULL;
    new_person->name = strdup(name);
    if (!new_person->name)
    {
        free(new_person);
        return NULL;
    }
    new_person->address = strdup(address);
    if (!new_person->address)
    {
        free(new_person->name);
        free(new_person);
        return NULL;
    }
    new_person->previous = NULL;
    return new_person;
}

person_t *destroy_person(person_t *p)
{
    person_t *previous = p->previous;
    free(p->name);
    free(p->address);
    free(p);
    return previous;
}
```

Note that the create_person() function checks whether there was enough memory available to duplicate the name and address of the new person, and takes appropriate action if there isn't. It will shortly become apparent why destroy_person() returns the previous field of the person_t to be destroyed.

Instead of an array or a pointer-to-the-first-element-of-a-dynamic-array, we put a pointer to the first element in the global scope as the 'root' of our database:

```
person_t *last_person = NULL;
```

It's called last_person because this variable will contain the last entry to be added to the database (and be NULL if there's nobody in the database).

Now, inserting a new entry in the database will consist of the following steps:

1. Create a new `person_t`,
2. Make its `previous` field point to the current last entry in the database,
3. Designate the newly created entry as the current last entry.

The following function fulfills these requirements:

```
int insert_person(const char *name, const char *address)
{
    person_t new_person = create_person(name, address);
    if (!new_person)
        return ERROR_OUT_OF_MEMORY;

    new_person->previous = last_person;
    last_person = new_person;

    return 0;
}
```

The two lines immediately before the `return` statement may look like a game of timblerig (or a shell game, if you're American), but they exactly implement the requirements number two and three of the list above.

After a few entries have been added in the database like this, there well be a 'chain' of entries in memory. In each entry, the `previous` field points to the entry added just before it. The figure below illustrates this:

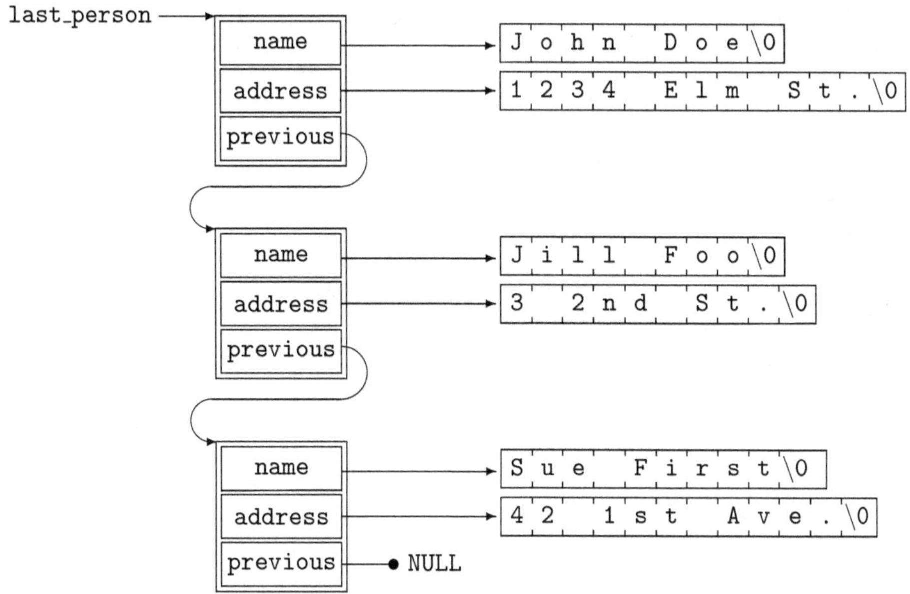

Note that some textbooks use '`next`' instead of '`previous`', along with '`first_person`' instead of '`last_person`'. This looks more sensible with the figure above, but you will then have to keep in mind that the list is filled in 'reverse order' (or the list is traversed completely each time a new `person_t` is added, or a separate pointer to the last element is kept around as well).

Finding entries in the database is slightly different from the version using an array, because to go from one entry to the next (or in this case, *previous*), the function needs to follow a *link:*

```
const person_t *find_person(const char *name)
{
    person_t *current = last_person;
    while (current)
    {
        if (!strcmp(current->name, name))
            return current;
        current = current->previous;
    }
    return NULL;
}
```

This function works by starting with the most recently added entry, checking whether it's the one we're looking for and returning it if so. Else, it will examine the previous entry, et cetera. The first entry added to the database will have its `previous` field set to `NULL`, so the function will end there (or immediately, if there was nothing added to the database yet).

So far, there has been little evidence presented that the version of the database using a linked list is superior to the one using a simple array. A first glimpse of the power of linked lists becomes apparent when we look at the function to remove entries from the database:

```
int remove_person(const char *name)
{
    person_t *previous = NULL;
    person_t *current = last_person;
    while (current)
    {
        if (!strcmp(current->name, name))
        {
            person_t *next = destroy_person(current);
            if (next)
                next->previous = previous;

            return 0;
        }
        previous = current;
        current = current->previous;
    }
    return ERROR_PERSON_NOT_FOUND;
}
```

Once a particular entry is found, it is a constant-time operation to remove it from the list – no matter how many entries there are in the list. This is because the entry to be removed is simply destroyed, and the link of the previous one is chained to the next one.

The same goes for insertions in the middle of the database. Suppose you want to keep the database in alphabetical order. In the case of a simple array, that means that whenever an entry needs to be inserted somewhere in the middle of the database, all entries following it need to be moved one spot 'to the right' to make room for the new entry (this is the exact opposite of what needs to happen when removing an entry from the database). In the case of a linked list, you can simply create the new entry and 'link it in' at the correct position. The only thing which needs to happen is the re-assignment of two pointers, instead of moving N entries (worst case).

On the other hand, looking things up in an array is faster. Therefore, the 'best' way of storing your data depends on several factors, one of the most important of which is the ratio between lookups and modifications.

If the number of lookups is much larger than the number of modifications (which is quite often the case with databases), it may well pay off to do some extra work during modifications to facilitate lookups. An obvious example is by keeping the database sorted.

Sorting entries in a list is a science on its own, but let us conclude this section by treating yet another way of storing our database, which makes use of alphabetical sorting.

Let us modify our database entry structure to include *two* links:

```
typedef struct person
{
    char *name;
    char *address;
    struct person *left;
    struct person *right;
} person_t;
```

(A modified `create_person()` function which sets the `left` and `right` entries to `NULL` is left as an exercise for the reader.)

The idea is to build up the database in the form of a *tree*, in which each *node* contains a single entry and two *branches*, each of which forms the root of a new tree. The idea is that the new tree 'growing from' the `left` pointer contains only entries with a `name` which is sorted *before* the current node, and the tree growing from the `right` pointer only entries with a `name` sorted *after* the current node.

It was said of the `strcmp` function ('string compare') that it takes two strings and returns zero if they are identical, and non-zero if not. In fact, `strcmp(a, b)` returns `1` if the string `a` is alphabetically *after* `b`, and `-1` if `a` is alphabetically *before* `b`. We can use this to insert new items into the tree. So far, we haven't really addressed the question as to what we should do when we find out that we're adding a person with the same name as someone else already in the database, but for now we'll just return an error code in that case:

```
#define ERROR_DUPLICATE_NAME (-4)

person_t *root = NULL;

int insert_person(const char *name, const char *address)
{
    person_t **p = &root;

    while (*p)
    {
        int position = strcmp(name, (*p)->name);
        if (position < 0)
            p = &((*p)->left);
        else if (position > 0)
            p = &((*p)->right);
        else
            return ERROR_DUPLICATE_NAME;
    }
    *p = create_person(name, address);
    return 0;
}
```

This function contains some rather cryptic pointer manipulations, so we'll go over it in detail. First, notice that the first line of the function declares p as a *pointer to a pointer to a* person_t, and it is assigned the *address of* the root of our database. Since root is in itself a pointer to a person_t, taking its address gives us a pointer to a pointer. These are relatively rare in C programs, but you're free to 'stack' as many pointer indirections as you like. We'll see why this extra indirection is useful in this case in a minute.

Next, we enter a similar loop as before with the linked list version of the database, and we're going to continue looping until we find an uninitialized person_t pointer (*i.e.*, one which is still set to NULL). The first time insert_person() is called, root will still be NULL, so we'll fall through the loop immediately. In that case, a new person_t is created and a pointer to it (returned from the create_person() function) is assigned to *p, *i.e.*, the memory p points to (in this case, the memory holding the value of root) is overwritten with this pointer to the newly created person_t.

The next time a person is added to the database, root will not be NULL anymore and the body of the loop is entered. The value of the new person's name is compared to the value of the current person_ts name, and the outcome is stored in the variable position.

If position < 0, that means that the new person has a name which alphabetically comes *before* the current one. In that case, the rather cryptic

```
p = &((*p)->left);
```

is executed, which simply means 'take what p points to (*p), which is a pointer to a person_t, then take its left entry, which is a pointer to a person_t again; finally take the address of that (so we get a pointer to a pointer to a person_t) and assign it to p'. More functionally described, we 'take a left turn' in the database tree, and make the left node the current one.

Similarly, we take the right node if the new name alphabetically comes *after* the current one. Lastly, like mentioned above: if neither is the case, the names must have been equal, and we return an error code.

Now, p points to a new node in the tree. If this node isn't used yet (*i.e.*, it is still NULL), the loop is terminated; else, the whole story repeats itself one level 'deeper' in the tree.

This is why the 'extra indirection' was handy; otherwise we would have to keep track of the previous node all the time (once the proper spot in the tree has been located, it is the *previous* node which needs to be modified to point at the new one).

An illustration of the result is in order. After the following few people have been added to the tree in order of appearance: Domenico, Arnold, Clara, Edward, Bela, and Francis, the tree structure will be composed like this (filled circles denote NULL pointers, *i.e.*, 'ends' of the structure):

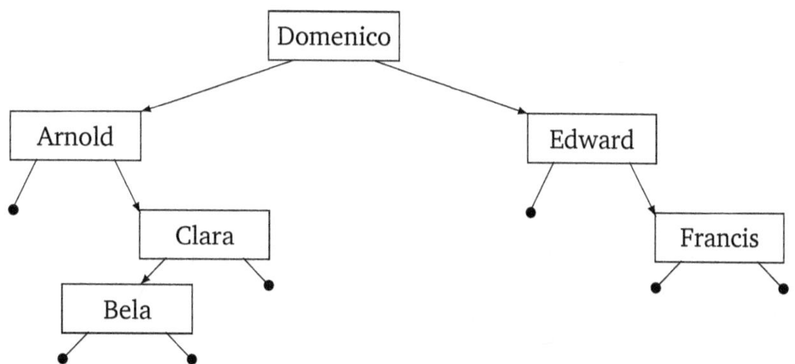

The nice thing about ordered trees is that it's generally faster to find entries in them. Let's look at the corresponding find_person() function:

```
const person_t *find_person(const char *name)
{
    person_t *current = root;
    while (current)
    {
        int position = strcmp(name, current->name);
        if (position < 0)
            current = current->left;
        else if (position > 0)
            current = current->right;
```

```
            else
                return current;
        }
        return NULL;
    }
```

To find Francis, only three nodes need to be examined, as opposed to six in the previous database structures. This reduces finding entries in the database from $\mathcal{O}(N)$ to $\mathcal{O}(\log N)$.

Trees have some interesting properties, and the questions and exercises at the end of this chapter deal with some of them.

6.4 Synopsis

C allows you to define new data types, either in the form of simple *aliases* for existing built-in types, or in the form of compound collections of *named, heterogeneous* data structures (as opposed to *numbered, homogeneous* arrays).

These data types (called `structs`) function almost like built-in types, in that they can be passed to functions as parameters, and be returned from functions as result values (but you cannot redefine *operators* to work with `structs` in C).

Just like with the built-in types, you can either pass `struct` parameters *by value*, in which case the callee receives a *copy* of the parameter, or *by name*, in which case only a *pointer* to the parameter is passed. This can be a performance benefit if the structures are relatively large.

C `structs` can be *nested, i.e.*, contain other `structs`, and can contain pointers to `structs` of their own type. This allows for 'linked containers' in which each element (or *node*) contains in addition to its data, one or more pointers to other nodes. Examples are *linked lists* and *trees*. In many cases, these structures are worthwhile alternatives to arrays.

6.5 Other Languages

The ability to define and use custom data types and structures has proven to be extremely useful, and most modern programming languages have either offered this feature from the onset (in Pascal, for example, they are part of the language under the name 'Records'), or have had them added in later revisions of the language (such as in BASIC).

During the history of computer programming, the emphasis in programming has shifted from the design of subroutines to the organization of data. First came the organization of programs in so-called *modules*, in which data types and associated functions are grouped together (a concept which C already supports, and which we will see more about in chapter 10).

When this concept is taken further, you make functions be a *part* of the data types they operate on; in this case, they are often called *member functions* or *methods*, and the types are usually called *classes*. This style of programming is called *object orientation* because the emphasis is not on 'functions performing certain tasks to solve a problem' (as in the 'recipe' style of procedural programming) but on 'objects representing certain entities from the problem domain' which can subsequently be manipulated.

In the database example of this chapter, there would be a 'database' class, offering functions to add, remove, or find objects of the 'person' class.

Having support for this paradigm in the programming language can simplify object oriented programming. A good example is the C++ programming language, which is not a pure object oriented language per se, but which was designed as a 'successor' to C, offering object orientation as one of several programming paradigms.

There, the problem of having to remember to initialize objects (such as the `person_t` data type) with the proper function, and certainly to 'destroy' them at the right time, is solved by making these constructor and destructor functions *part of the class*. They are automatically called whenever an object is instantiated or leaves scope, respectively.

Most of the 'modern' programming languages either offer objected oriented design as one of the supported paradigms (such as C++, Visual BASIC, and Delphi), or 'enforce' it by forbidding the definition of functions outside classes (such as Java, C#, and Eiffel).

Object orientation includes other features such as *inheritance*, where a class can *derive from* another class; this means that the derived class extends the functionality of the 'base class' (the class it derives from) and can be used wherever an object of the base type is required. The canonical example is a hierarchy of animals: There would be an `animal` base class with derived classes `dog` and `cat`; whenever a function expects an object of type `animal` you are free to pass it either a `dog` or a `cat`. This works only one way though: When a function wants a `cat`, you can't pass just any `animal` to it. (Note that this is not the same as the 'parameter conversion' in C programs, where if you provide an `int` argument to a function expecting a `double`, it is implicitly converted.)

Another important aspect is *polymorphism*, which means that 'derived' objects can be manipulated through pointers to their base class – as if the objects had multiple types. For example, each `animal` would have an `eat()` function (which `dog` and `cat` implement slightly differently); which particular implementation is chosen depends on the 'real' type of the `animal`. Of course, only objects of type `dog` would have a `bark()` function. You could further derive several breeds of dogs, which all implement this `bark()` function in their own way, etc.

Some object-oriented programming languages rely on 'run-time polymorphism' which means that type information is only deduced while the program is running (as opposed to during compilation), so continuing the same example you'd have a function simply taking an 'object', and then querying the object for its properties at run-time ('if the object is a `dog`, see if it has a `bark()` function and call it'). Examples of such languages are SmallTalk, one of the pioneering languages in object orientation, and Objective-C, which adds run-time polymorphism to C. The latter is mainly used on the Apple Mac OS X platform.

There is more to object-oriented programming of course, but a full treatment of is outside the scope of this book.

6.6 Questions and Exercises

6.1 Implement a 'real program' using the database functionality from §6.3.4. It should offer a simple text-based interface which has 'add', 'remove', and 'find' commands.

6.2 Make different versions of the database program using the alternative storage schemes presented in §6.3.5. Measure their relative performance using a pre-generated list of (random) names; first for filling the database with these names, for looking up names, and for removing them. It may be interesting to use different test sets and to plot the results against the database size. Is one scheme always faster than the others? Does one scheme lend itself better for larger databases? If so, what is the critical size?

6.3 In the tree graph on page 143, where would a new entry for 'Gabriel' end up? And for 'Camille'?

6.4 What would the tree structure look like when all names were added to the database in alphabetical order? What would that do to the efficiency of looking up names in it afterwards?

6.5 ⋆⋆ The requirement 'for best performance, names should be added to the database in randomized order' will impose quite a limit to the usefulness of the database program (see the previous question). Therefore, when the tree becomes too 'unbalanced' (*i.e.*, there are far more 'left turns' than 'right turns' or vice versa), it needs to be 'balanced' again. First, add functionality to score the 'balance' of the tree; then, implement a function to re-balance it.

Chapter 7
Files

For most men life is a search for the proper manila envelope
in which to get themselves filed.

Clifton Fadiman

So far, we have limited ourselves to user input from the keyboard, which is one step up from simply typing the relevant information right into your program, and output to the screen. If you want your program output to be more 'permanent', you can write it to a *file*. It is equally possible to read data from files instead of from the keyboard.

7.1 Storage Devices

Historically, files were stored on magnetic tape. This is still used in backup systems. The predominant type of persistent storage now is the *hard disk*, which contains one or more spinning magnetic disks (with rotational speeds varying from 4200 rpm in laptops to 15 000 rpm in high-end server disks). Typical capacity at the time of writing is several hundred gigabytes.

Another popular type of storage are *optical disks* such as CD-ROMs and DVDs. The capacity of a CD-ROM is approximately 700 megabytes and DVDs can hold several gigabytes.

The venerable *floppy disk* (so called because the original 8" and 5¼" disks were in a flexible envelope) is rapidly becoming extinct. The usual capacity of the current, 3½" version is 1.44 MB, which by todays standards is not very large; they are also not very durable or reliable.

The place of floppy disks is being taken by *flash memory* which is solid-state. There are various embodiments of flash memory, such as CompactFlash which has the same interface as an IDE hard disk (see 1.2.2), various types of memory cards often used in digital photo cameras, and 'pen drives' or 'USB sticks' which can be plugged directly into an USB port of the computer.

7.2 File Systems

Most operating systems offer support for a *file system*, (some even for various file systems), with which is meant a system for the structured storage of data on a persistent storage device, such as a hard disk. 'Persistent' means that the data is kept even when the device is switched off. Although there are examples of file systems in 'volatile' memory, these are considered exceptional.

The job of a file system is to allow storage and retrieval of data in files, and organizing these files. Usually, files are *named,* and in the vast majority of file systems they can be organized in *directories* or *folders*. The complete sequence of directories followed by the name of a file is called the *path* to a file.

On Windows systems, directory names are separated by backslashes, so a file called `database.c` in the directory `programs` which in itself is stored in the `My Documents` directory would have a path looking like `C:\Documents and Settings\john\My Documents\programs\database.c`. The '`C:`' at the beginning of the path is the *drive letter;* on Windows, each disk drive has its own drive letter. This is an inheritance from DOS[1].

On most other operating systems, the directory separator is a forward slash, and drive letters are not used. Instead, the file system can span multiple physical disks and these disks can be *mounted* anywhere in the file system. On a typical UNIX system, the above path would look like `/home/john/programs/database.c`.

Most operating systems have a concept of a *current working directory* for a program, which is the directory you were 'in' when the program was started. The paths in the examples above were *absolute paths*, but you can usually also specify *relative paths* which are relative from the current working directory. Suppose you were in the `My Documents` directory on Windows or in your home directory on a UNIX system, the `database.c` file would have a relative path of `programs/database.c`.

It is also possible that the operating system supports *network file systems*, *i.e.*, present a file system physically located on a different computer somewhere on the network as if it were a local file system.

7.3 File I/O in C

The C standard library offers support for working with files by abstracting away much of the details of the underlying operating system. You can create files, open and close them, and read and write data. The only point of attention is the directory separator; the backslash is a 'special character' in C (namely, the *escape character*) so to specify a path on a Windows system, you either need to type '\\' for each backslash in the path, or simply use forward slashes. The latter is supported under Windows as well, and increases the 'portability' (ease of transporting to a different platform) of your program, making it less Windows-specific.

[1] More precisely, each *partition* has its own drive letter. Partitioning a disk means to split it up in several 'virtual' disks.

Let's look at an example of a program which writes something to a file.

```
#include <stdio.h>

int main(void)
{
    FILE *fp = fopen("hello.txt", "w");
    if (!fp)
    {
        printf("Could not open hello.txt for writing\n");
        return 1;
    }
    fprintf(fp, "Hello, world!\n");
    fclose(fp);
    return 0;
}
```

After running this program, there should be a file called `hello.txt` in the current working directory, containing the text 'Hello, world!'. Let's go through the example line by line.

First, note that we include the `stdio.h` header file, which contains all the necessary definitions and prototypes to perform file I/O.

The `fopen()` call is used to open a file; the first parameter is the file name (in this case, a relative path, so it will be located in the current working directory); the second parameter designates 'how' to open the file. In this case, a `"w"` is specified, meaning 'open the file for writing'. If the file does not already exist, a new one is created; if it does already exist, it is truncated and its contents overwritten(!) by the program. It returns a `FILE` pointer, or `NULL` if it failed to open the file. In that case, the file name specified may be invalid (because it contains characters which are illegal in file names on this particular file system), or you lack permissions to create files in this particular directory, or the file may already exist and you lack the permissions to truncate and/or overwrite it. When opening a file for reading by specifying `"r"` instead of `"w"`, it may also mean that the file is simply not there.

You can also specify `"a"` to open the file in 'append' mode; everything you then write to the file is appended to it.

The `FILE` pointer can subsequently be used to write to the file. In this case, we used the `fprintf()` function, which is just like `printf()` except that its first parameter is a `FILE` pointer designating the file to write to. In fact, `printf(...)` is just shorthand for `fprintf(stdout, ...)`. `stdout` is one of the three 'default' files which every C program has access to. The other two are `stdin` for standard input, and `stderr`, which is usually used to print error info to, so as to separate this from 'normal' program output.

After we're done with the file, we need to `fclose()` it, after which the `FILE` pointer can no longer be used. This call also 'flushes' everything written to the file to the disk. The system is free to 'buffer' data because it is more efficient to write data to disk in large blocks instead of in many small amounts. Flushing means that all buffered data is written to disk.

The next example reads lines from a file and prints them:

```
#include <stdio.h>

int main(int argc, char *argv[])
{
    FILE *fp;
    if (argc != 2)
    {
        fprintf(stderr, "Usage: %s <filename>\n", argv[0]);
        return 1;
    }
    fp = fopen(argv[1], "r");
    if (!fp)
    {
        fprintf(stderr, "Couldn't open %s\n", argv[1]);
        return 2;
    }
    while (!feof(fp))
    {
        char line[80];
        fgets(line, 80, fp);
        printf("%s", line);
    }
    fclose(fp);
    return 0;
}
```

First, we check whether a single argument is specified on the command line, in which case `argc` is 2 (remember that `argv[0]` contains the program name itself). If there isn't, note that we print the error message to `stderr`. This is good style, because especially programs like this are often used to redirect their output to a different file (see §7.6). If we just used `printf()` here, the error message would end up in that file as well.

Next, we try to open this file for reading, checking whether it succeeded.

Then, we enter a loop; the `feof()` function checks whether the end of the specified file has been reached. While this is not the case, we read one line worth of characters from the file. If a line in the file happens to be longer than 79 characters (`fgets()` makes sure there is room for the terminating zero), it will continue with the line in the next iteration of the loop. Then, the line is printed to the standard output, and the loop is continued.

7.4 Structure of Data in Files

7.4.1 Human-Readable Files

There are various ways of storing data in files. The most straightforward one from a user perspective is the *human readable format*. For example let's consider a program which reads a list of numbers from a file and prints out their average value. The numbers are to be typed in the file with a text editor, which means they are stored as *strings*, and need to be converted into numbers by the program. This is similar to what we saw in in the 'factor example' in §5.9. In the code below, we've skipped the error handling of non-existent files, assuming the file exists and is accessable.

```
#include <stdio.h>

int main(int argc, char *argv[])
{
    int n = 0;
    double sum = 0;
    double value;

    FILE *fp = fopen(argv[1], "r");
    while (fscanf(fp, "%lf", &value) == 1)
    {
        sum += value;
        n++;
    }
    fclose(fp);

    if (n > 0)
        printf("Average is %f\n", sum/n);

    return 0;
}
```

As you may have guessed, `fscanf()` is the equivalent of the `scanf()` function we've seen before, but scanning the specified file instead of the standard input. In fact, `scanf(...)` is shorthand for `fscanf(stdin, ...)`. The `fscanf()` function in the program above takes care of parsing the (human readable) number into a `double`. The return value of `fscanf()` is the number of entries read, so at the end of the file (or whenever an invalid character is found) it will return zero.

Remember that the format string used by `(f)scanf` is similar to that used for `printf()` (see the table on page 53), except that to read a `double`, you need to specify `%lf` ('ell-eff', as opposed to just `%f` for a `float`).

An alternative would have been to use `fgets()` to get a line from the file, and use `atof()` to do the conversion to a floating point number. The drawback there is that this presumes that the numbers are on separate lines, while the `fscanf()` variant can also handle numbers separated by whitespace.

7.4.2 Binary Files

The main advantage of human-readable (*i.e.*, text) files is that they are, well, human-readable. You can look at their contents without any special viewer software, and you can use any text editor to edit them. The drawback is that to convert their contents into something a program can work with, like numbers, the computer needs to parse these strings. This leads to a performance overhead. Usually, this overhead is negligible compared to the time it takes to actually read information from disk, but in the case of large data structures, it may become noticeable. Also, there is a substantial file size penalty.

Especially for data structures which are produced and consumed by a computer program only (*i.e.*, which no human being is supposed to look at directly, let alone modify), you can also read and write data by using the representation of the data in memory directly. For this purpose, you can use the `fread()` and `fwrite()` functions:

```
size_t fwrite(const void *ptr, size_t size, size_t nmemb, FILE *fp);
size_t fread(void *ptr, size_t size, size_t nmemb, FILE *fp);
```

The function `fwrite()` writes `nmemb` objects, located at `ptr`, each of `size` bytes, to the file `fp`. The function `fread()` reads `nmemb` objects, each of `size` bytes, storing them in memory at `ptr`.

To use these functions, you must open the file in 'binary mode' (as opposed to 'text mode') by using `"rb"` and `"wb"` in the `fopen()` function. This prevents any 'smartness' in the library, for instance converting different 'newline' designators. This is mostly for historical reasons; on many systems the `"b"` flag is simply ignored and files are always opened in binary mode.

An example:

```
typedef struct
{
    double x;
    double y;
    double z;
} vector3;

int main(void)
{
    FILE *fp;
    vector3 array[10];

    calculate_vectors(array, 10);

    fp = fopen("vectors.dat", "wb");
    fwrite(array, sizeof(vector3), 10, fp);
    fclose(fp);
    return 0;
}
```

What needs to be realized when using binary files is that they are not inherently portable across different machines. When a 16-bit `short` is stored in memory on a certain system, there are two ways of doing so. The two bytes making up the 16-bit number are stored in two adjacent memory locations, one being the *least significant byte* (LSB) and the other the *most significant byte* (MSB). The actual 16-bit value is $256 \times \text{MSB} + \text{LSB}$. On some systems, the LSB is stored first, in the memory location with the lowest address. These systems are called *little endian*; Intel x86 systems are an example. On other (*big endian*) systems, like the PowerPC, the MSB comes first. A similar story goes for 32-bit integers.

Now, this is not something you usually notice or care about, because the way a system stores its numbers in memory is something entirely internal to the system – after all, it's the *same* system which reads them back.

However, if you write this data to disk, transport it to a machine with a different 'endianness', and read it back into memory, the resulting number is typically vastly different from the original!

Another thing which may cause surprises, even on the same system, is *padding*. Consider the following `struct`:

```
struct example
{
    char   a;
    short  b;
    short  c;
    int    d;
};
```

On many systems, the CPU can access data fastest if it is *aligned* in memory, *i.e.*, a 32-bit integer is stored on a memory address which is a multiple of 4. On some systems, this alignment is not just an optimization, the system simply *can't* access the data otherwise.

This means that an instance of `struct example` may not look like this in memory:

a	b_0 b_1	c_0 c_1	d_0 d_1 d_2 d_3

but probably more like this:

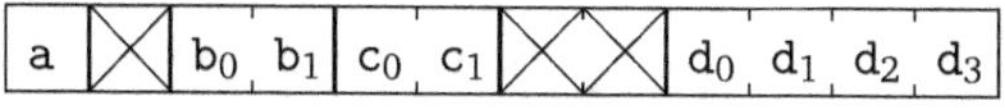

The crossed-out bytes are not used, and are inserted there by the compiler as *padding*. So you may expect that `sizeof(struct example)` is 9, but in reality it is probably 12. This is to ensure that `b` and `c` start on an even address, and `d` starts on a multiple of 4.

You can imagine that writing a data structure to disk from a system with a certain padding, and reading it back again on a system with different padding leads to the wrong results. This may be the *same* physical system: sometimes it is a compiler option whether and how to pack structures.

7.4.3 File Formats

We've seen that there are several ways of storing information in files. For many types of data, there are standard ways in which they are stored, so that files can be written and read by any computer program conforming to these standards. There are many examples of standard *file formats*.[2]

There are file formats for (bitmap) images, such as JPEG, PNG, GIF, TIFF, BMP, etc.; for vector oriented graphics, such as PostScript and SVG files; for audio files, such as MP3, AIFF, and WAV; for three-dimensional structures, for word-processing files or typeset documents, for hyperlinked text files, for programming languages, for executable files on various computer systems, for generic structured data, etc.

Most file formats start with a *header* containing a 'magic word' to designate the file format, followed by some information about the file. In the case of a bitmap image file, this header would contain information about the dimensions of the image, the type of color space, etc. In some cases, the actual content of the file is a fairly straight-forward mapping of the desired data structure in memory; in other cases, the file contents may have been *encoded* – for example, *compressed* to save storage space or bandwidth (as is the case with JPEG and MP3).

In the former case, it is usually easy to write code for reading and writing such files yourself; in the latter case, there are often *libraries* (see chapter 12) which offer functionality to do so. In chapter 11, we'll write a function for saving BMP files.

Some file formats are not 'open', in the sense that they are not defined or described outside the company that uses them; a notorious example are many word processing document file formats, which have to be 'reverse engineered' if you want to read their contents on a platform not officially supported; a practice which is even illegal in some countries.

It is also worth noticing that some file formats are covered by *patents* in some part of the world, including the USA, and that even though the file format is well-described, you have to pay royalties to program an encoder or decoder for it.

7.4.4 Random Access

So far, we've only looked at reading and writing data *sequentially*. When one `fread()` is finished, the next `fread()` will continue at the position in the file where the previous one left off. This 'current position' is nicely hidden behind the `FILE` pointer, and you normally don't need to worry about it. You can even open a file multiple times; each one receives a new `FILE` pointer. Reading from it through these different `FILE` pointers lets you read from different positions in the file.

There are a few functions available for 'repositioning' this current file position, and finding out where in the file the next read or write would take place. The most common functions are:

[2]'The nice thing about standards is that there are so many to choose from' – *Andrew Tanenbaum*

```
int fseek(FILE *fp, long offset, int origin);
long ftell(FILE *fp);
```

An `fseek()` sets the 'current position' in the file at `offset`, counted in bytes from `origin`. This `origin` may be one of `SEEK_SET` (count from the beginning of the file), `SEEK_CUR` (count from the current position), or `SEEK_END` (count from the end of the file). In other words, `fseek(fp, 0, SEEK_SET)` puts the current position at the beginning of the file. A shortcut for this is `rewind(fp)`, which is a remnant of the old, tape-based storage systems.

The function `ftell()` returns the current position in the file, so the following piece of code can be used to determine the size in bytes of a given file:

```
long len;
fseek(file, 0, SEEK_END);
len = ftell(fp);
```

Note that 64-bit file systems are common even on otherwise 32-bit systems, because the limit for file sizes in a 32-bit file system is 4 gigabytes (the maximum number of bytes addressable by a 32-bit integer). With the advent of multi-media capabilities on personal computers, this limit was rapidly becoming a problem. For example, one hour of full-size video footage from a DV camera takes up over 13 gigabytes. BeOS was one of the first personal computer operating systems using a 64-bit file system, but most systems offer support for this now. Note that this means that the `offset` parameter in `fseek()` and the return value of `ftell()` on such systems cannot be ordinary `long`s; often the prototype of these functions uses a special typedef `off_t` which maps to a 64-bit integer type.

Also, the `size_t` parameters in `fread()` and `fwrite()` are usually 64-bit in these systems. Of course, on a pure 32-bit system, you wouldn't be able to read in a block of data of which the size wouldn't fit in a 32-bit integer, because you couldn't address enough memory to put it in.

7.5 A Useful Example: Least Squares Fitting

In this section, we will develop a program to make a linear fit through a set of data points using the 'least squares' method. The data points are to be given in a text file.

7.5.1 Theory

Given a set of points in the *xy*-plane (usually some kind of measurements) which are supposed to lie on a straight line, the goal is to find the equation for the line which 'best fits' the points. 'Best fit' means that the sum of the squared vertical differences of each data point from this line is minimal; hence the name 'least squares fit'.

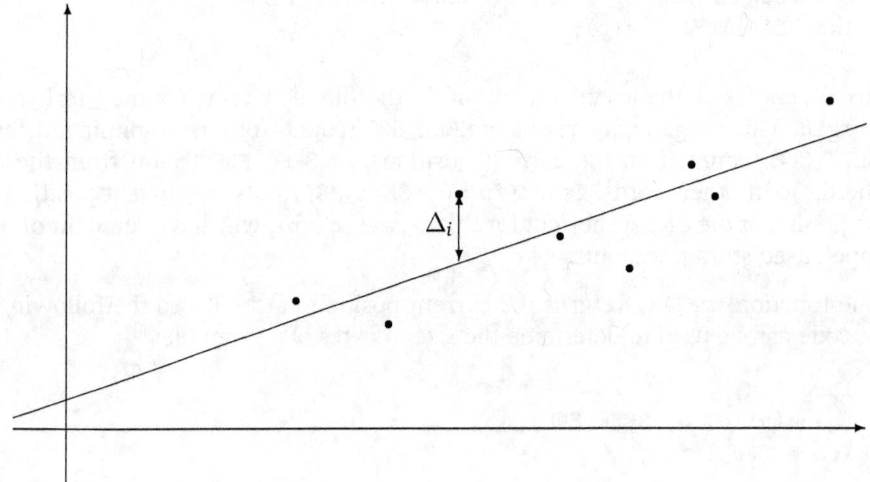

If we denote the ith data point by (x_i, y_i) and the straight line by $y = a_1 x + a_0$, then the difference Δ_i for this point is given by

$$\Delta_i = y_i - (a_1 x_i + a_0),$$

and the sum of all squared differences (usually called χ^2, 'chi squared') is

$$\sum_{i=0}^{n-1} \Delta_i^2 = \sum_{i=0}^{n-1} \big(y_i - (a_1 x_i + a_0)\big)^2.$$

We now have to find a_0 and a_1 such that this sum of squares is minimal. This leads to

$$\begin{aligned} a_0 n + a_1 \sum_i x_i &= \sum_i y_i \\ a_0 \sum_i x_i + a_1 \sum_i x_i^2 &= \sum_i x_i y_i \end{aligned}$$

with solutions

$$\begin{aligned} a_1 &= \frac{n \sum_i x_i y_i - \sum_i x_i \sum_i y_i}{n \sum_i x_i^2 - \left(\sum_i x_i\right)^2} \\ a_0 &= \frac{1}{n} \sum_i y_i - \frac{a_1}{n} \sum_i x_i. \end{aligned}$$

7.5.2 Implementation

The first thing we need is a function to read a set of data points from a file. This isn't as straight-forward as it seems, because we don't know in advance how many data points the file will contain, and hence we cannot allocate an array 'big enough' beforehand.

Therefore, we cannot make a function which will 'fill up' a *given* array with data from the file. It is best to make a function which will *create* this array for us, remembering that we will have to `free()` this array ourselves when we're done with it.

Also, we need to know how many data points were read from the file, so this function needs to return this information somehow. However, any function can only return a single type. We have two choices to solve this problem: Either we define a new data type (a `struct`) which holds both the array and its size, or we pass the function a pointer to an `int` as one of its parameters, and have it fill in the number of points read through that pointer. We'll choose the latter approach here, which is the 'out value' approach from §5.6.

First, a simple `typedef` for each data point:

```
typedef struct
{
    double x;
    double y;
} point_t;
```

Next, a function to create an array of `point_t` and fill it with the contents of a given file. We'll first present the code, and then walk through it:

```
point_t *create_from_file(int *num_read, FILE *fp)
{
    int n = 0;
    int current_size = 8;     /* Just a guess */
    point_t *array;

    /* Some sanity checks */
    if (!num_read)
        return 0;
    if (!fp)
        return 0;

    array = malloc(current_size*sizeof(point_t));
    while (fscanf(fp, "%lf, %lf", &array[n].x, &array[n].y) == 2)
    {
        n++;
        if (n == current_size)
        {
            current_size *= 1.5;
            array = realloc(array, current_size*sizeof(point_t));
        }
    }
    if (!feof(fp))
        fprintf(stderr, "Malformed input at entry %d\n", n);
```

```
    *num_read = n;
    return array;
}
```

The function starts by making a guess about the number of data points in the file. Here, we chose 8, but any number greater than one will do. Next, it performs some sanity checks: if the `int` pointer passed to the function was `NULL`, we will crash if we try to write through it, so we make sure that isn't the case. Also, if we're passed an invalid `FILE*`, we return immediately.

Once these initial sanity checks are passed, it allocates enough room for an array of `current_size point_ts` and enters a loop to read the data from the file.
In each iteration, we use `fscanf()` to read two `doubles` from the file, separated by a comma, storing them in the current position in the array (which is `n`). `fscanf()` returns the number of items it successfully scanned, and we check whether indeed two numbers have been read. As long as that is the case, we increment the current index. If this reaches `current_size`, that means the next entry in the file will not fit in the current array, so we increase the size of the array by a factor of 1.5 and re-allocate it using the `realloc()` function (see page 134). For brevity, this example doesn't check for an out-of-memory condition.

If we weren't able to read two numbers, we may have simply reached the end of the file. If not, there must have been something wrong with the layout of the numbers in the file, so we print out a warning message. Most likely, a comma was forgotten; in that case, `fscanf` can successfully read the first number, but then looks for a comma and can't find it, so it will return 1 instead of 2.

The last two lines of the function are to write the number of data points read from the file through the `num_read` pointer, and return the current array to the caller.

Next, we need a function to perform the actual least squares fitting. An example of such a function is given below. It takes a `point_t` array with given size and two `double` pointers, into which the resulting a_0 and a_1 are written (as defined in the Theory subsection above).

```
int least_squares(point_t p[], int n, double *a0, double *a1)
{
    double Sx, Sy, Sxx, Sxy;
    int i;
    Sx = Sy = Sxx = Sxy = 0;
    if (!p || n < 2 || !a0 || !a1)
        return 1;

    for (i = 0; i < n; i++)
    {
        Sx += p[i].x;
        Sy += p[i].y;
        Sxx += p[i].x*p[i].x;
        Sxy += p[i].x*p[i].y;
    }
```

```
        *a1 = (n*Sxy - Sx*Sy)/(n*Sxx - Sx*Sx);
        *a0 = Sy/n - (*a1)*Sx/n;
        return 0;
    }
```

Looking at the formulas for a_0 and a_1 on page 156, you see that you need to iterate over all points in the array. The crux of the `least_squares()` function above is therefore a single `for`-loop.

Before this loop is entered, a set of accumulating variables is defined and initialized to zero, via a common C idiom of 'chaining' them all together in a single assignment.

Next, a quick sanity check is performed to see whether all the parameters passed to the function were valid, to prevent from crashing by writing through a `NULL` pointer, or dividing by zero. Then, the loop is entered and the respective sums are calculated (`Sx` stands for $\sum_i x_i$, `Sxy` for $\sum_i x_i y_i$, etc.)

Finally, the values for a_0 and a_1 are calculated and written to the respective pointers. Note that a_1 needs to be calculated first because its value is used in the calculation of a_0, and that its value needs to be obtained from *dereferencing* the pointer `a1`. Remember from Section 5.5 that if `pa` is a pointer to a `double`, then `*pa` represents 'what `pa` points to'.

7.6 Redirection

Although not a feature of the C programming language per se (it's rather a feature of the operating system), we'll shortly mention *redirection* here because it's a powerful concept which can be quite useful.

It has been mentioned before that the functions `printf()` and `scanf()`, for example, are really shortcuts for `fprintf()` and `fscanf()` with the `stdout` and `stdin` parameters pre-filled in, respectively. The nice part about `stdin` and `stdout` is that most operating systems allow *redirecting* these.

For example, given the basic 'Hello, World!' program we started this book with, and assuming it's called '`hello`', then when you run it from the command line by simply entering

```
hello
```

you would get the expected output on the computer screen. However, if you type:

```
hello > file.txt
```

(assuming the file `file.txt` doesn't already exist), the output of the program is written to `file.txt` instead. The way this works is that usually, `stdout` is linked to the computer screen, but by telling the operating system you want to *redirect* this output to a file instead, the underlying system `fopen()`s this file for you (behind the scenes) and points `stdout` to it before your program starts. That way, your program is writing to a file without even knowing it.

Similarly, you can redirect its *input*. For example, take the program to calculate the sum of a list of numbers typed by the user (see page 4.4). You could also run this by typing something like

```
sum < numberlist.txt
```

after which the data read by `scanf` from `stdin` is in fact read from the file `numberlist.txt`.

You can even link the output of one program to the input of another by typing

```
numgen | sum
```

assuming that `numgen` is some kind of program which generates a list of numbers terminated by a single zero (because `sum` happens to depend on that). This starts the programs `numgen` and `sum` simultaneously, and whatever `numgen` writes to its `stdout` appears on `stdin` for `sum`. This is called *piping*. A program which reads input from `stdin` and writes processed data to `stdout` is called a *filter*. You could imagine a program, for instance, which reads text from `stdin`, changes certain characters into certain other characters (say, to change all upper case characters into lower case), and writes the resulting text to `stdout`. The next chapter will show some examples of filters turning plaintext into encrypted text, and vice versa.

To make your programs work with piping and redirection, stick to a few simple guidelines:

1. Don't assume you can 'rewind' your input using `fseek()`; make sure you code your application so as to expect 'live' input from a user

2. When your program needs to read input from a file, make sure to offer an option to read from `stdin` instead:

   ```
   int main(int argc, char *argv[])
   {
       FILE *in;
       if (argc < 2)     /* no file name has been provided */
           in = stdin;   /* so use the standard input instead */
       else
           in = fopen(argv[1], "r");
       read_from_file(in);
       if (in != stdin)
           fclose(in);
       return 0;
   }
   ```

3. Conversely, if your program expects a file name as one of its commandline parameters to write its output to, allow running it without such a filename, writing to `stdout` instead.

4. It's better to read up to 'end-of-file' instead of up to a magic number, such as our `sum` program did. The user can enter 'end-of-file' when typing numbers manually by entering Control-D on UNIX systems or Control-Z on Windows. Otherwise, your program might get 'stuck' waiting for the magic number when it's piped to the output from another program which doesn't know about that convention.

7.7 Synopsis

You can make your programs read from and write to *files* instead of the keyboard and screen, respectively. A *file* is a named, persistent collection of data. Linking the name to the stored data is the job of the *file system*, which is part of the operating system and which usually provides helpful features such as structured hierarchical storage of files (sometimes much more, such as automatic versioning of files, network transparent access, etc.).

The file handling routines in C are largely platform-agnostic; the name of the file is passed as a C string and the standard library offers functions to read and write from and to files based on a platform-independent `FILE` pointer.

Data is stored in files in some *format*, which can be *text-based* or *binary*; the former is easier to work with if the data has to be modified 'manually' because it can be in a 'human readable' format; the latter is more efficient in terms of storage space and/or performance. C provides functions to operate on files in both ways; there are functions available to convert human readable input to native data types.

In the case of binary files, proper caution needs to be taken because the internal representation of data types is system and compiler dependent.

On many operating systems, including Windows, you can *redirect* the standard input or output from programs, including connecting programs together in a sort of 'pipeline', where the output of one program is fed into the input of the next one.

7.8 Other Languages

Most other programming languages offer support for reading from and writing to files. In some, like old BASIC dialects, the support is not very extensive, especially on home computer systems which offered only crude forms of persistent storage. They often stored data on audio cassette tape.

Some programming languages impose 'structured storage' where a file is structured much like a C `struct` (or a collection of those), whereas C offers a very 'low-level' view of a file as being just a stream of bytes, the interpretation of which is left to the programmer.

Some file formats can be nicely described in an object-oriented manner, especially files which can describe a structured, hierarchical collection of 'data entities'. Working with those can be more 'natural' in an object-oriented programming language.

7.9 Questions and Exercises

7.1 Derive the least squares formulas in §7.5.1.

7.2 Use the functions from §7.5.2 to make a program which takes a filename as a commandline argument, reads an array of x, y points from the named file, and prints out the line equation for the best fitting line through the data points.

7.3 Find a solution for a second-order fit, *i.e.*, where $y = a_2x^2 + a_1x + a_0$, and implement it.

7.4 Write functions to convert 16-bit integers and 32-bit integers from big-endian to little-endian, and vice versa. Hint: It is possible to cast a pointer-to-`int` to a pointer-to-`char` and treat it as an array of 4 `chars`.

7.5 Write a program which takes one command line argument N and prints out the numbers 1 through N, each on a separate line. Next, modify the program to calculate the average value of a given list of numbers on page 151 to allow reading from `stdin`, and set up a command line piping the output of the number generator program into the averaging program. Verify that the average value of N consecutive numbers $1, 2, \ldots, N$ is $N(N + 1)/2$.

Chapter 8
Bits and Shifts

There are 10 kinds of people in the world – those who understand binary, and those who don't.

Unknown

In §1.3, it was explained how computers use the binary system to represent numbers. In this chapter, we will look at some features of the C language for manipulating bits directly. We will then look at an example in which we will use this 'bit-fiddling' to solve an interesting puzzle, and after an intermezzo on optimization conclude with an application of (very) basic cryptography.

8.1 Bitwise Operators

One bit (short for 'binary digit') is the smallest amount of information digital computers can work with. A bit can only have one of two possible values: zero or one. A group of eight bits[1] is called a *byte*. Half a byte (*i.e.*, four bits) is sometimes called a *nybble*.

Normally, you are not concerned much with how the computer uses groups of bits to represent numbers (or even other symbols, such as characters in a string), unless you are running into limits of either size or precision. However, sometimes it can be useful to manipulate the bits of a certain value directly. The four most-often used operations known from 'binary logic', namely 'and', 'or', 'exclusive or', and 'not', are available in C as the so-called 'bitwise operators' `&`, `|`, `^`, and `~`. These may look familiar from the section about boolean expressions (4.3), with the distinction that these operators work on all the bits in their operands simultaneously. We'll take a look at all these operators in turn.

The simplest operator is the 'not' operator, which toggles each bit in its operand: each 0 is replaced by a 1 and vice versa. If an `unsigned char a` has the value of 00110101 in binary notation (*i.e.*, 53 in decimal), then `~a` has the binary value 11001010 (*i.e.*, 204 in decimal). If you remember from §1.3.1, you can use this to represent negative values: if `a` is a positive integer, then `~a + 1 == -a`.

[1]As was mentioned in §1.2.2, there *are* systems with a different number of bits in a byte. These are, however, exceedingly rare and mainly of historical interest.

The other three operators take two operands. The 'and' operator places a 1 for each resulting bit if the corresponding bits of its operands are both 1. The 'or' operator places a 1 if either operand has a 1 in the corresponding position (or both operands do), and the 'exclusive or' (sometimes called 'xor') operator places a 1 if *exactly* one of its operands has a 1 in the corresponding position.

In tabular form:

```
  1100          1100          1100
  1010 &        1010 |        1010 ^
  ----          ----          ----
  1000          1110          0110
```

The 'and' and 'or' operators are often used to test, set, and clear specific bits in a value. For example, to test whether the 0th (*i.e.*, rightmost) bit in a variable is set, you could do:

```
int is_odd(int a)
{
    return a & 1;
}
```

(Note that an integer value is always odd if its rightmost bit is set, and always even if its rightmost bit is not set.) Selecting only certain bits this way is also called *masking*.

To make sure bits 2 and 4 of some variable x are set (for whatever reason), you could do

```
x = x | 20;  /* 20 decimal is 00010100 binary */
```

Or, since just like with the 'normal' mathematical operators, the bitwise operators can be combined with assignment:

```
x |= 20;
```

In combination with the 'not' operator, it is easy to clear the designated bits in a variable, since you can simply mask all the *other* bits:

```
x &= ~b;    /* clear all bits in x which are set in b */
```

This first toggles all the bits in b, so what was a 1 becomes a 0 and vice versa; then it masks x with the resulting value, in effect clearing all bits which were set in b.

A second group of bitwise operators is formed by the *shift operators*, which move all the bits in a value left or right by a certain number of positions: `x << n` shifts all the bits in x left by n positions, and `x >> n` shifts all the bits in x right by n positions.

When bits to the left, the bits at the right (*i.e.*, the least significant bits) are filled with zeros. Since 'adding a zero to the right' in effect doubles a value (just like 'adding a zero to the right' in the decimal representation is equivalent to 'multiplying by ten'), if x is a positive integer, `x << n` is the same as $x \times 2^n$ (provided this value is still representable in the resulting type, *i.e.*, it is not too big). By the way, if you couldn't find the answer to question 1.2 (page 27) before, perhaps now you can.

Shifting to the right is similar. If x is a positive integer, then x >> n is simply $x \times 2^{-n}$, rounded down. That is because the least significant bits simply 'fall off', without any mathematical rounding taking place.

However, if x is a *negative* value, the result of shifting it is 'implementation defined' (even in the C99 standard). This is because some CPUs have instructions for *logical shifting* (*i.e.*, adding zeros regardless of the sign of the operand), whereas others have instructions for *arithmetic shifting*, where the 'sign bit' is taken into account. Many CPUs even have both. C doesn't impose limits on the CPU in this regard, and you should avoid using shifts on negative values.

Using shifts in combination with the other bitwise operators, you can define functions to set, clear, and test specific, numbered bits in a value:

```
void clear_bit(int *x, int n)
{
    *x &= ~(1 << n);
}

void set_bit(int *x, int n)
{
    *x |= 1 << n;
}

int is_set(int x, int n)
{
    return x & (1 << n);
}
```

Note that in the first two functions, we need to pass a *pointer* to x, because the function needs to modify its value. In code which does a lot of 'bit-fiddling', you often see functions like these implemented as preprocessor macros instead:

```
#define CLEAR_BIT(x,n) ((x) &= ~(1 << (n)))
#define SET_BIT(x,n)   ((x) |=   1 << (n))
#define IS_SET(x,n)    ((x) &   (1 << (n)))
```

8.2 3D Puzzles

8.2.1 Introduction

In the early nineties of the previous century, a certain type of toy gained immense popularity. They were known under a variety of names ('Happy Cubes', 'Wirrel Warrel Cubes', 'Snuzzles', or 'Snafooz' in the USA). Invented by Dirk Laureyssens and featured in the pavilion of Belgium in the EXPO'92 in Sevilla, they became 'toy of the year' in several countries.

They basically consist of a small 'mat' of foam rubber, about half a centimeter thick, in which six pieces are punched. You can take out these pieces and combine them into a cube. Each of the pieces has six 'notches' or 'holes' on each side (puzzles with different numbers are available too, but we'll focus on six). Each puzzle piece therefore looks somewhat like the picture below:

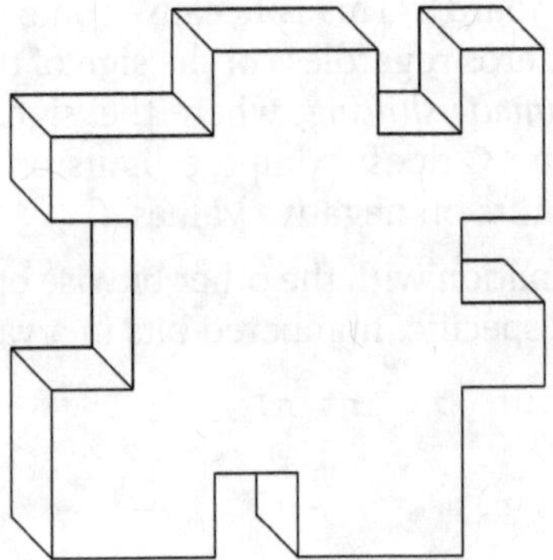

Depending on the level of difficulty of the puzzle, there are more or fewer ways to combine the pieces into a cube. The finished cube would look somewhat like this:

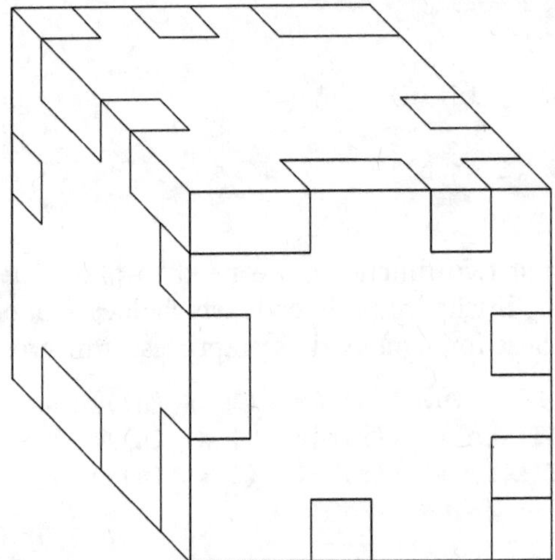

We will spend the remainder of this section on a program to solve these puzzles, using binary arithmetic to decide whether a certain configuration 'fits'. We can use binary arithmetic here by designating a 'hole' by a zero and a 'notch' by a one; two sides then fit if they yield a binary pattern of six ones when 'xor'ed together. We will see that some special measures are needed to take the corners into account, but the basic idea will prove useful.

This type of puzzle-solving is not just for fun. In computational biochemistry, for example, one problem comparable in nature (but not in complexity, of course) is to determine computationally what the three-dimensional structure of certain large molecules (say, proteins) looks like, or whether and how enzymes bind to certain substrates.

8.2.2 Solving Strategy

The number of possible ways to combine six pieces into a cube is enormous. First, there are 6! = 720 ways to place six pieces on the sides of the cube (the number of *permutations* of the six pieces), but each of these pieces can be rotated in four ways and be flipped over and rotated four times again – yielding a total of 8 orientations for each piece. That means that for each permutation, there are $8^6 = 262\,144$ possible configurations, yielding a total of 188 743 680.

Now, we have counted several configurations which lead to the 'same' cube multiple times. After all, when we take an assembled cube and rotate it a bit, we shouldn't count this as a 'different configuration'. Since there are 24 ways of orienting a cube (each of the sides in a certain direction, and four 90° rotations of each orientation) and we can turn each configuration 'inside out' as well, we 'only' have $188\,743\,680/48 = 3\,932\,160$ really different configurations.

If these numbers aren't immediately obvious, it may help to picture a 'flattened' cube as in the illustration below. (The dotted arrows point out which 'corner notches' end up in which 'corner holes'.)

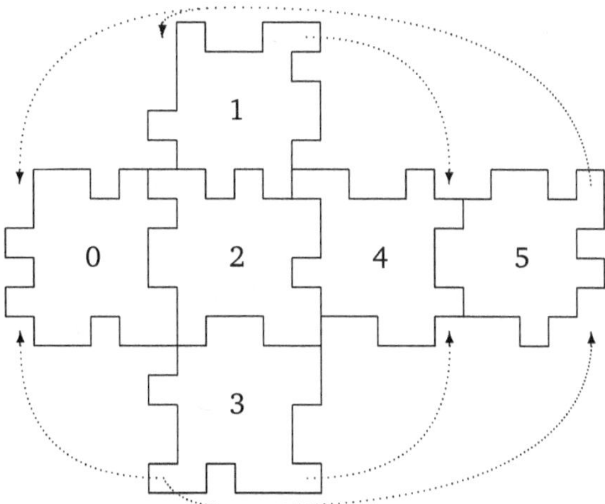

If you fold up this cross into a cube, then topple over the cube so a different side ends up on the bottom, and then unfold it again into a cross with the same layout as in the picture above, then this 'new configuration' has the pieces arranged somewhat differently. This configuration would be one of the 188 743 680, but it's clearly not a 'different' solution to the puzzle. Also, flipping the entire cross over and folding it up would yield an 'inside out' cube which also fits, and which arguably isn't a 'different' solution either.

It is, however, not entirely trivial to figure out whether two configurations are 'equivalent', so let's first make an estimate of whether it'll be doable to simply try them *all*, in an 'extremely brute force' approach.

Let's estimate the amount of time it takes to generate and test one single configuration to be below one microsecond. This is reasonable, because in this 1 μs a modern CPU can execute at least in the order of magnitude of one thousand instructions. Testing the 12 rims and eight corners shouldn't take more than a hundred instructions (mostly simple 'xor' instructions and some additions, so no complex calculations), leaving plenty to generate the configuration.

This would give an upper limit of some three minutes to run through all configurations. This is within reason. Remember that we're only after a rough estimate here – if we ended up with a much larger time needed, we'd have to think of a smarter way to generate the correct configurations. As it is, we can continue with the 'brute force approach'.

8.2.3 Permutations

Saying *how many* permutations of a given number of items there are is one thing, *generating* them all is more involved. If you're asked to write down all permutations of the four digits 0, 1, 2, and 3, you'll probably do so in a certain order, and not by just thinking up random permutations until you've found all 24 (otherwise, you are probably not the intended audience for this book). If you do so, you'll probably notice that they ended up in ascending numerical order – 'by accident'. In fact, most people with a 'scientific inclination' will generate the list column-by-column, starting by writing down six zeros on separate lines, then six ones, etc., then moving to the second column and writing the permutations of the *remaining* digits, etc. This is, in fact, classic *recursion* at work (see §4.9).

We could, of course, use a recursive algorithm to generate all the permutations of the puzzle pieces. Another approach is to find an algorithm which, given a certain permutation, yields the 'next' one.

Let's look at part of a list of permutations of the numbers 0 – 4, and suppose the permutation pointed to by the arrow is the 'current' one:

$$\begin{array}{ll} \vdots & \\ 12340 & \\ 12403 & \\ 12430 & \longleftarrow \\ 13024 & \\ 13042 & \\ \vdots & \end{array}$$

The algorithm to find the next permutation is:

1. Starting from the right, look for a number which is less than the one to its right. This is the number which will be swapped with the next higher one. In the example above, this would be the 2.

2. Swap the number found in step 1 with the next higher one to its right. In the example above, this would be the 3, yielding the intermediate result of 13420.

3. Finally, sort the numbers to its right in ascending numerical order again. Note that at this point, they are always in *reverse* order. This would yield 13024, which is the next permutation.

The following function does just that: given an array of n integers which are in a certain permutation, it calculates the next.

```
#include <limits.h>    /* for INT_MAX, see below */

void next_permutation(int a[], int n)
{
    int i;

    /* Find the first number (starting from the right)
       which is less than the one to its right. */
    for (i = n - 2; i >= 0; i--)
    {
        if (a[i] < a[i + 1])
        {
            /* Now, find the next higher number (i.e., the
               smallest number larger than a[i]) to its right */
            int m = INT_MAX;
            int k = i;
            int j;
            for (j = i + 1; j < n; j++)
            {
                if (a[j] > a[i] && a[j] < m)
                {
                    m = a[j];
                    k = j;
                }
            }
            swap(&a[i], &a[k]);

            /* Reverse the remaining numbers */
            for (j = 1; j <= (n - i)/2; j++)
                swap(&a[i + j], &a[n - j]);

            return;
        }
    }
}
```

INT_MAX is a constant (defined in limits.h) which is equal to the largest possible value which an int can hold on this particular platform. It is a standard trick when looking for some minimum to initialize with INT_MAX, because all other values are *at most* equal to this.

The function swap() is a simple 'helper function' which can be defined as follows:

```
void swap(int *a, int *b)
{
    int tmp = *a;
    *a = *b;
    *b = tmp;
}
```

Note how it takes two *pointers* to integers, and exchanges the values they point to via a temporary value.

8.2.4 Data Type and Manipulation

To hold the description of one puzzle piece, we'll use a `struct` holding the binary representations of each of its four sides:

```
typedef unsigned char side_t;

typedef struct
{
    side_t top;
    side_t bottom;
    side_t left;
    side_t right;
} piece_t;
```

Since we're examining puzzles with six positions on each side, an `unsigned char` should be a good data type for each `side_t`. This is a `typedef` 'for future convenience': On page 181 we will change its definition in pursuit of a (slightly) more efficient implementation.

Each of the `side_t`s will hold a binary representation of a side, where the `top` and `bottom` ones will have their least significant bits on the right side of the piece (as is consistent with how you would write down the ones and zeros above and below the puzzle piece) and the `left` and `right` ones will have them at the top of the piece (*i.e.*, at their natural positions after rotating the piece 90° clockwise). All in all, the binary representation of the example piece on page 166 would be `top` = 001101b, `bottom` = 110110b, `left` = 110010b, and `right` = 001011b, or 13, 54, 50, and 11 in decimal, respectively. Note that this way of encoding is slightly redundant, because each corner spot is encoded twice.

Functions for rotating and flipping these puzzle pieces are also needed. For these, we will also need a function to reverse the six bits:

```
side_t reverse6(side_t s)
{
    int j;
    side_t r = 0;
    for (j = 0; j < 6; j++)
    {
        r <<= 1;
        r |= (s & 1);
        s >>= 1;
    }
    return r;
}
```

This function adds new bits to the right of r as they are shifted out of s. In more detail, it makes room for a new 'right' bit in r by shifting the current contents one bit to the left, then puts the current 'right' bit of s in that position, and finally pushes that bit out of s by shifting its contents one bit to the right.

Using this reverse6() function, we can write a rot90() to rotate a puzzle piece 90° counterclockwise, and a flip() function which flips it around its vertical axis:

```
void flip(piece_t *p)
{
    side_t tmp = p->left;
    p->left = p->right;
    p->right = tmp;
    p->top = reverse6(p->top);
    p->bottom = reverse6(p->bottom);
}

void rot90(piece_t *p)
{
    side_t tmp = p->left;
    p->left = p->top;
    p->top = reverse6(p->right);
    p->right = p->bottom;
    p->bottom = reverse6(tmp);
}
```

Since we chose to pursue a non-recursive algorithm, we will also need a function to generate the configurations (meaning the orientations of the individual pieces) *within* each permutation. We can do this by keeping track of these configurations in an array of integers, one for each piece, where the numbers 0 – 3 designate the four different rotations with one face up and 4 – 7 with the other face up:

```
int next_configuration(piece_t p[], int cfg[], int n)
{
    int pos;
```

```
        for (pos = n - 1; pos >= 0; pos--)
        {
            if (cfg[pos] == 7)  /* Had all rotations for this piece */
            {
                flip(&p[pos]);  /* Restore to its original state... */
                cfg[pos] = 0;
                continue;       /* ... and continue with next piece */
            }

            if (cfg[pos] == 3)  /* Rotated four times? */
                flip(&p[pos]);  /* Then it's time for the other side */
            else
                rot90(&p[pos]);

            cfg[pos]++;
            return 1;           /* Found a new configuration */
        }
        return 0;               /* No more configurations left */
    }
```

8.2.5 Fitting It All Together

Given functions to generate permutations and configurations, the actual program calling these functions is relatively straightforward. A walk-through of the program follows below.

```
#include <time.h>

int main(int argc, char *argv[])
{
    piece_t pieces[] = { { 20, 50, 54, 22 },
                         { 36, 38, 59, 12 },
                         { 13, 54, 50, 11 },
                         {  8, 13, 20, 50 },
                         {  9, 18, 24, 19 },
                         { 12, 11,  8, 44 } };
    int permutation[] = { 0, 1, 2, 3, 4, 5 };
    int m = fac(6);
    int i;
    int configs = 0;
    int solutions = 0;
    clock_t start = clock();

    for (i = 0; i < m; i++)
    {
        int c[] = { 0, 0, 0, 0, 0, 0 };
        piece_t config[6];
```

```
            int j;
            for (j = 0; j < 6; j++)
                config[j] = pieces[permutation[j]];

            do
            {
                configs++;
                if (configuration_fits(config))
                {
                    double took = (double)(clock() - start)
                                  /CLOCKS_PER_SEC;
                    printf("Checked %d configs, (%g sec/config)\n",
                           configs, took/configs);
                    print_configuration(config, permutation, c);
                    solutions++;
                }
            } while (next_configuration(config, c, 6));

            next_permutation(permutation, 6);
        }
        printf("Found %d solutions\n", solutions);

        return 0;
    }
```

The first statement of the program initializes the array of puzzle pieces with an example puzzle (this is the encoding of the puzzle on page 167). In the array `permutation` the current permutation is kept. The `fac()` function in the next line is the 'factorial' function as developed in exercise 4.3 – you can also simply fill in '720' here. We'll also keep track of the number of configurations already tested, the number of solutions found so far, and the time it took to check each configuration (so we can verify our assumption that this would be $< 1\,\mu s$).

Then, the 'permutation loop' is started. For each permutation, the array `config` is initialized with the pieces in the current permutation. The array `c` is used to keep track of the current configuration, to be passed to the `next_configuration()` function. Once this is all set up, the 'configuration loop' is started. The `configuration_fits()` determines whether the current configuration can be folded up to a 3D cube (we'll take a look at such a function below). If it does fit, the program prints out how many configurations have been checked so far, and how long it took on average for each configuration. A 'fitting' configuration is also printed out – the implementation for `print_configuration()` will also be examined below.

After all configurations for a certain permutation have been generated, `next_configuration()` will return 0 so the 'configuration loop' is exited, and the next permutation will be generated.

To define a `configuration_fits()` function, all vertices of the cube need to be examined. The binary representations for each side of the vertex, 'xor'ed together, need to yield the pattern 111111. Special care needs to be taken for the corners of the cube, because three pieces need to be taken into account there. Looking at the flattened puzzle on page 167, for example, you can see that the vertex between piece 0 and piece 2 is entirely determined by those two pieces, whereas the vertex between piece 4 and 5 is completed by the top right notch of piece 1 and the bottom right notch of piece 3 (as indicated by the dotted arrows).

Let's first define a few handy macros to make the huge boolean expression we're about to construct a little more readable:

```
#define LT(p) (p.top >> 5)
#define RT(p) (p.top & 1)
#define LB(p) (p.bottom >> 5)
#define RB(p) (p.bottom & 1)
```

Using these, and keeping an eye on the layout on page 167 to see which vertices are connected, and which corner notches are to be taken into account, we arrive at the following horrendous expression:

```
int configuration_fits(piece_t p[])
{
    return (((p[0].left^p[5].right)+(p[3].left & 32)+LT(p[1]) == 63)
         && ((p[0].bottom^p[3].left)+(p[5].right & 32)+LB(p[2]) == 63)
         && ((p[0].top^reverse6(p[1].left))+LT(p[2])+(RT(p[5]) << 5) == 63)
         && ((p[0].right^p[2].left)+(p[3].top & 32)+LB(p[1]) == 63)
         && ((p[3].bottom^reverse6(p[5].bottom))+(p[0].bottom & 32)+RB(p[4]) == 63)
         && ((p[3].top^p[2].bottom)+(p[0].right & 32)+LB(p[4]) == 63)
         && ((p[2].top^p[1].bottom)+(RT(p[0]) << 5)+LT(p[4]) == 63)
         && ((p[1].top^reverse6(p[5].top))+(p[0].top & 32)+RT(p[4]) == 63)
         && ((p[4].bottom^reverse6(p[3].right))+(p[2].right & 32)+LB(p[5]) == 63)
         && ((p[2].right^p[4].left)+(RT(p[3]) << 5)+RB(p[1]) == 63)
         && ((p[4].top^p[1].right)+(RT(p[2]) << 5)+LT(p[5]) == 63)
         && ((p[4].right^p[5].left)+(p[3].right & 32)+RT(p[1]) == 63));
}
```

(A smaller font size had to be applied and 'cosmetic' spaces had to be removed to get each part of this expression on a single line.) Each part of the expression consists of the 'xor' of two adjacent sides, plus the relevant corner notches, shifted to the right position if needed. The result must be 111111 (63 in decimal) for each part. Note how some of the sides are passed through the `reverse6()` function defined earlier, because they are 'flipped over' when folding up the cube.

This function returns 1 if *all* sides match exactly. When such a configuration is found, it is printed out to the screen. However, printing out a configuration in a more-or-less graphical way, so that it resembles a flattened cube layout, is not a trivial matter. C makes very few assumptions on the kind of computer screen used to display its output – in fact, as we saw in the previous chapter, it need not be a screen at all. Basically, the 'lowest common denominator' is a typewriter-like device in which characters can be printed on a line, and the whole screen scrolls up by one line when it's full. Imagine typing out a puzzle layout this way, using, say, asterisks to represent the notches, and spaces to represent holes. Imagine that you are not allowed to move the print position upwards or backwards (*i.e.*, you are not allowed to use the backspace key on your typewriter, and are not allowed to rotate the paper knob to move the paper downwards). Quite a complicated task, if all you're given is a representation of the layout in a `piece_t` array.

Of course, many computer screens (or terminals) do offer functionality to reposition the cursor (using special so-called *escape sequences*), and there are libraries (see chapter 12) which present a common set of C functions to use those. Otherwise, programs like full-screen interactive text editors would be horribly inefficient. Since we aren't trying to make an interactive program, however, there is a simple trick we can use: Simply take a two-dimensional character array of appropriate size, use that to 'draw' our layout into (being able to address all positions freely in whichever order we like), and then print out this array to the screen in one go once it's ready.

A set of functions to print out a configuration using this technique is given below. As added bonus, they will also 'label' each puzzle piece with its original number, and even tell you whether it has been flipped over. This way, if you use your program to solve a 'real' puzzle, numbering the puzzle pieces as you encode them into the program, you can use the output of the program easily to piece together the solution.

The first function sets up such a 'display buffer' of the appropriate size (20×27 characters in our case) which has room for one extra space between all the puzzle pieces. It then calls a `print_piece()` function (defined below) to draw each puzzle piece at the appropriate position in the layout. It takes the permutation array `a` as an extra parameter and places the numbers of the puzzle pieces in the display array. This is done by adding the character code for the character '0' to the number values in the array – using the fact that on the vast majority of computer systems, the human-readable representations for the digits 0 – 9 have consecutive number values (see §5.3.1). Also, the current configuration array `cfg` is used to determine whether a certain piece has been flipped over (in which case its configuration number is > 3) and a character 'B' is added if that is the case. Finally, the temporary display buffer is written to the real screen.

```
void print_configuration(piece_t p[], int a[], int cfg[])
{
    char display[20][27];
    int i, j;
    for (i = 0; i < 20; i++)
        for (j = 0; j < 27; j++)
            display[i][j] = ' ';

    print_piece(display, &p[0], 7, 0);
    print_piece(display, &p[1], 0, 7);
    print_piece(display, &p[2], 7, 7);
    print_piece(display, &p[3], 14, 7);
    print_piece(display, &p[4], 7, 14);
    print_piece(display, &p[5], 7, 21);

    /* Label the pieces */
    display[9][2] = a[0] + '0';
    display[2][9] = a[1] + '0';
    display[9][9] = a[2] + '0';
```

```
        display[16][9] = a[3] + '0';
        display[9][16] = a[4] + '0';
        display[9][23] = a[5] + '0';

        /* Mark whether they are flipped over */
        if (cfg[0] > 3) display[9][3] = 'B';
        if (cfg[1] > 3) display[2][10] = 'B';
        if (cfg[2] > 3) display[9][10] = 'B';
        if (cfg[3] > 3) display[16][10] = 'B';
        if (cfg[4] > 3) display[9][17] = 'B';
        if (cfg[5] > 3) display[9][24] = 'B';

        /* Output the temporary display buffer to the real screen */
        for (i = 0; i < 20; i++)
        {
            for (j = 0; j < 27; j++)
                printf("%c", display[i][j]);
            printf("\n");
        }
    }
```

The `print_piece()` function draws a single puzzle piece into the temporary display buffer with its top left corner at the specified row and column position:

```
    void print_piece(char d[20][27], const piece_t *p, int row, int col)
    {
        int i, j;
        for (i = 0; i < 6; i++)
            if (p->top & (32 >> i))
                d[row][col + i] = '*';

        for (i = 1; i < 5; i++)
        {
            if (IS_SET(p->left, i))
                d[row + i][col] = '*';

            for (j = 1; j < 5; j++)
                d[row + i][col + j] = '*';

            if (IS_SET(p->right, i))
                d[row + i][col + 5] = '*';
        }

        for (i = 0; i < 6; i++)
            if (p->bottom & (32 >> i))
                d[row + 5][col + i] = '*';
    }
```

This uses the macro defined on page 165. The output of a single configuration would look something like this:

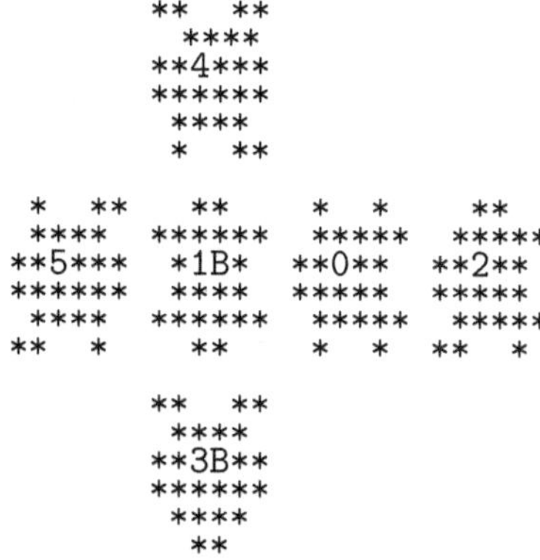

Incidentally, this is not a solution of the puzzle on page 167. That one only has a single solution, meaning that in each of the 48 equivalent 'solutions' found by the program presented here, either none of the pieces are flipped over, or all of them are.

8.3 Intermezzo: Optimization

The program presented in the previous section is horribly inefficient. A human solving this kind of 3D puzzles would take a totally different approach. If you already know that two pieces won't fit together, you would never generate new configurations by fiddling with some other pieces *elsewhere* in the puzzle, since that would never make the puzzle as a whole suddenly fit. Yet this is precisely what the program does: Even if the first two pieces in the configuration do not match, it will still generate *all* the configurations with those two pieces in that position.

Instead, you would probably start out by finding two pieces which match, then gradually expand by finding a piece from the remaining set which matches somewhere in the solution you are building up. If you can't find a piece that matches anywhere, you'll have to remove one of the pieces already connected and see if you can find a different fitting configuration.

Implementing a solution using this (recursive) technique is left as an exercise for the reader (exercise 8.3 to be precise), and such a solution will easily be *orders of magnitude faster* than the brute force approach taken in the previous section. Just like we saw in section 4.8, if you want to speed up a program the first thing to look at is the algorithm used.

However, suppose a certain algorithm is the best you've got. There are some tricks and techniques you can apply to optimize your program so that it runs faster, and we will examine a few in this section. They have nothing to do with binary arithmetic, but the program presented in this chapter is the first one with 'sufficient complexity' that optimization makes sense.

8.3.1 Compiler Switches

The first optimization can be done without changing even a single line of code in your program, and only providing some specific instructions to the *compiler*. Most compilers can produce code with various levels of optimization, and the default settings are usually 'unoptimized'. This is because switching on optimization usually results in both longer compile times and larger executable files. If you are willing to make this trade-off, you can make the compiler work a little harder on your program.

Don't expect any miracles, but in programs involving a huge number of relatively simple calculations, you can get performance increases ranging from a few percent to almost a factor of two (if you're very lucky). With GCC, you can specify the `-O3` command line switch (for **O**ptimization level three) and with Microsoft Visual C++, you can specify `/Ox` to switch on full optimization.

Things a compiler might do include checking whether a certain computed value is used at all, or whether a certain piece of code has any effects. Sometimes, people write code like

```
for (i = 0; i < 1000000; i++)
    ;
```

and time how long it takes a certain computer to perform this 'empty loop' as a measure of how fast this computer is. With (aggressive) optimization turned on, a compiler might notice that this loop is 'useless' and simply remove it altogether.

A second thing influencing the code a compiler generates is being more specific about the kind of processor you are targeting. Especially in the Intel (x86) world (including the various Pentiums and Core processors, and x86-compatible offerings from competitors such as AMD), there are many different types of processors which vary not only by clock speed, but also by availability of certain special-purpose CPU instructions (see §1.2.1) or specifics regarding how best to generate code for them.

Most compilers are 'conservative' in their default assumptions about the processor. This is nice for when you have to ship your compiled executable program to different people (for instance, you may be able to make the assumption that everyone interested in your program has *at least* a Pentium), but it's not-so-nice if your compiler generates code which is optimal for a lowly 386 and you will only ever run it on your own Core 2 machine.

With GCC, you can specify the `-march=`*cpu-type* and `-mcpu=`*cpu-type* compiler options, and Microsoft Visual C++has a set of `/G`*n* options, with *n* a number. For example, `/G3` optimizes for 80386, `/G5` for Pentium, and `/G7` for Pentium 4 or Athlon systems.

Note that when you switch on such compiler switches, the resulting executable will no longer run on 'lower' systems.

8.3.2 Avoiding Calculations

Of course, the fastest calculations are the ones you don't do. The compiler can only do so much in this respect, even when optimizing aggressively, because it can't (easily) be sure that a function call will always have the same result. For example, if you would have code like

```
for (i = 0; i < 1000000; i++)
    x += sin(y);
```

it would make sense to calculate the value of `sin(y)` *once*, multiply it by a million, and add it to `x`. However, if the code looked like this:

```
for (i = 0; i < 1000000; i++)
    x += random();
```

(see §3.6) then the compiler had better not conclude that the function call to `random()` does not depend on `i`, and hence needed to be called only once.

Let us first introduce a new keyword, `static`. We'll see how this is useful in a minute. If you declare a variable `static` within a function, then this variable will retain its value through subsequent calls to that function. For example:

```
int count(void)
{
    static int n = 0;
    return n++;
}
```

The first time `count()` is called, `n` is initialized to 0. It is then returned and incremented to 1. The next time `count()` is called, `n` will still have the value 1, so this is returned, and `n` is incremented again, etc. This can be used for checking how many times a function is called, which can be handy when you are trying to improve the performance of a complex program. Another example is a function returning the current average of all the numbers passed to the function so far:

```
double running_avg(double x)
{
    static double sum = 0;
    static int n = 0;

    sum += x;
    n++;
    return sum/n;
}
```

One interesting candidate for optimization in our 3D puzzle program is the `reverse6` function (see page 171). Using the 'count' trick above, it is easy to see that it is called more than 400 million times over the course of a single program run. Making it only one nanosecond faster would already have a measurable impact on the total performance.

What makes it so interesting is that of those 400 million function calls, there are only ever 64 possible input-output pairs. Instead of calculating the bit-reverse of the parameter at each function call, it would be better to calculate them all *once* and store them in some kind of table. The nice thing is that with the help of `static`, this can be done without the callers of the function even noticing it (apart from it being faster, of course):

```
side_t reverse6(side_t s)
{
    static int initialized = 0;
    static side_t lut[64];
    if (!initialized)
    {
        int i;
        for (i = 0; i < 64; i++)
        {
            int j;
            side_t a = i;
            side_t b = 0;
            for (j = 0; j < 6; j++)
            {
                b <<= 1;
                b |= (a & 1);
                a >>= 1;
            }
            lut[i] = b;
        }
        initialized = 1;
    }
    return lut[s];
}
```

In this function, a `static` table called `lut` (for 'lookup table') is filled with all possible answers the first time the function is called. At each following call, `initialized` will be 1, so the table won't be filled again. This simple optimization results in a speed increase of about 30% (!).

The 'LUT trick' can be applied when the number of possible input-output pairs is small compared to the number of calls made to the function, and when it is not a problem that the first call to the function takes significantly longer than the rest.

This implementation combines two often-used techniques that seem fundamentally incompatible: That of *lazy evaluation* (calculate a value only when you really have to, in order to prevent calculating something which turns out not to be needed) with *eager evaluation* (calculate *more* than you have to, in order to save time in the future).

8.3.3 Minor Optimizations

There are a few other techniques which can squeeze out a few more percent of performance. This is usually not worth the hassle, but sometimes a feature of the underlying computer architecture shows through which may be interesting to note.

Unfortunately, what may be beneficial for one particular computer architecture may make no difference on another, or even have a detrimental effect. If this kind of performance tweaking is really necessary, you will have to perform some experimenting on the exact processor and architecture your program is going to run on. One of the few examples where a few extra percent can make the difference, is when you are doing 'real time' calculations: data is coming in at a fixed rate (say, from some kind of measurement device) and you need to perform some kind of processing on it. Even if your processing code is only a few percent slower than the rate at which the data comes in, you will have to somehow store all the data and perform the processing off-line.

The most important optimization is to take the effects of cache memory into account (see §1.2.2). If you access memory in patterns which are 'expected' by the cache memory, *i.e.*, in a linear fashion, it will be most efficient. For example, if you are processing a large, two-dimensional array of data, it is often (quite a bit) faster to access it row-by-row instead of column-by-column. That is because in the former case, you are accessing adjacent memory locations, whereas in the latter case, you are making big 'jumps' through memory. What can happen then is that the cache memory has just been filled with a fresh 'burst' of data from main memory, only to find out that the next memory access is many bytes further away. By the time you are ready to process the second column, the data from the first row has already been flushed from the cache.

Another thing to note is that some data types are more 'natural' to a certain CPU than others. For example, many CPUs are most efficient at accessing memory 32 (or even 64) bits at the time. If you access single bytes, the processor will have to retrieve the whole surrounding 32-bit 'word' of memory, modify the single byte, and write the word back. In the 3D puzzle example, this is visible when we play with the `typedef` for the `side_t` type. In this particular program, an `unsigned char` is more than big enough to hold the data for a single vertex of the puzzle, and in §8.2.4 it was `typedef`'d as such. The reason this was a `typedef` in the first place was to facilitate experimenting with different data types. On a Pentium 4, for instance, the resulting program is about 5% faster if this `typedef` is changed to an `int` instead (although you might have expected it to be slower, since it has to operate on *more* data now).

Moving to an even more esoteric optimization, it is necessary to go a bit deeper into features of modern CPUs. In order to get CPUs with ever more performance, CPU designers have added *pipelines* to the processor. In such a pipeline, one instruction can be decoded while the next is already being fetched from memory. These pipelines have become deeper and deeper, and processors also no longer process single instructions one by one. Instead, they can even 're-order' instructions by noting that one instruction isn't dependent on the output of another, and decide that it can process these instructions in parallel. At each clock tick, a new instruction enters the pipeline, and all others move on to another stage.

One problem in this design is what to do with 'branches' in the code (*i.e.*, when the program has to jump elsewhere in the code to continue), and especially *conditional branches*. The compiler generates such branches every time it finds an `if`-statement, `for`-loop, etc. The problem is that when filling up the pipeline with the 'next' instruction, it depends on the outcome of, say, a comparison, *which* statement is the next.

Most CPUs simply pick one of the possible outcomes (for example, they always follow the branch taken when the condition is met) and continue filling up the pipeline as if this was the case. This is called 'branch prediction' (even if the CPU always 'predicts' the same outcome). If, once the outcome of the conditional branch is known at a later stage in the pipeline, it turns out that the pipeline was filled with the *wrong* code path, the entire pipeline contents are just 'thrown away', with an obvious performance penalty.

The `reverse6()` function in the 3D puzzle is a good example. One compiler structured the code such that moving a shortcut *above* the LUT-filling code, *i.e.*,

```
side_t reverse6(side_t s)
{
    static int initialized = 0;
    static side_t lut[64];
    if (initialized)
        return lut[s];
    else
    {
        /* fill the lut */
        ...
        return lut[s];
    }
}
```

resulted in a program which was 7% faster than with the version on page 180. Apparently, the generated code made the processor falsely assume that the LUT-filling code was the path most likely to be taken. Interestingly though, with some other (more recent) compilers, there was no difference in performance. Compilers are getting better and better at detecting situations like these, and tuning the generated code to take advantage of the specific properties of a certain CPU. That also means that it is getting less and less necessary to spend your time on these minor optimizations. A good rule of thumb is 'first make it work, then make it work fast'.

8.4 (Very) Basic Cryptography

The desire to hide or encrypt messages so that information can be shared between certain people without being accessible to others has been with us for thousands of years. In this section, we will look at two (basic) encryption algorithms, one of which features the 'xor' operator.

8.4.1 The Caesar Shift Cipher

In his treatise *De Bello Gallico*, Julius Caesar mentions the use of cryptography to send messages to allied troops, so that even when intercepted, the messages would make no sense to the enemy. One of the schemes he employed was to replace each letter in the message by a letter in the alphabet a few positions further on. The simplest version of such an encryption method would replace the letter A by a B, the B by a C, etc., and the Z by the A. In that method, *'A secret message'* would become *'B tfdsfu ndttbhf'*. A shift by two positions would replace A by C, B by D, etc., so that the message would become *'C ugetgv oguucig'*. The recipient only needs to know by how many positions to shift back the letters in the alphabet to decipher the message. This type of cipher is known as a 'Caesar shift cipher', or simply a 'Caesar shift'.

A special case is the shift by 13 positions. Since our alphabet has 26 characters, you can use the same algorithm both to *encrypt* and to *decrypt* (because shifting by 13 twice effectively cancels the encryption). This extremely simple shift is known as a *'rot13 encoding'* in computer science. 'Rot' stands for 'rotation', which is the proper term for a shift where the elements 'falling off' on one side are re-inserted at the other side.

A rot13 encrypting/decrypting is extremely simple to implement:

```
#include <stdio.h>
#include <ctype.h>

int main(void)
{
    int c;
    while ((c = getchar()) != EOF)
    {
        if (islower(c))
            c = (c - 'a' + 13) % 26 + 'a';
        else if (isupper(c))
            c = (c - 'A' + 13) % 26 + 'A';
        putchar(c);
    }
    return 0;
}
```

The `getchar()` function[2] reads a single character from `stdin`, and `putchar()` writes a single character to `stdout`. If there is no more input available on `stdin`, `getchar()` returns `EOF`. In the `while` loop, you can see a compact C idiom for assignment and comparison. The variable `c` is assigned the return value of `getchar()`, and it is immediately compared with `EOF`. This is a handy notation, because the value of `c` is needed inside the body of the `while`-loop.

The `islower()` and `isupper()` functions return whether their argument represents and lower case or upper case letter, respectively. These functions are declared in the `ctype.h` header. This particular rot13 implementation leaves all other symbols (including numbers) as-is. When it is determined that the character read from `stdin` is, in fact, a letter, its numerical value is 'normalized' by subtracting the numerical value of the first letter of the alphabet. That means the numerical value of 'a' and 'A' becomes zero, the value of 'b' and 'B' becomes 1, etc. This is so we can then simply

[2] On many platforms, `getchar()` is actually a macro

add 13 modulo 26 (which takes care of the 'wrap-around'). Finally, the numerical value is positioned back so it represents a letter again (see §5.3.1).

Caesar-shifted messages are extremely simple to break. Since there are only 25 possible shifts, you would simply take (part of) the message and apply brute force until sensible text appears. Rot13 is still widely used though, but only to protect from 'accidental' viewing. Some email and net-news readers have rot13 filters built-in; you could publish a review of a movie but rot13-encode the paragraph in which you give away the ending, to prevent spoiling the movie for a reader accidentally reading it.

8.4.2 The Vigenère Cipher

Since the Caesar shift is so easy to break, cryptographers came up with a different encryption method. Instead of simply shifting each letter a fixed number of positions in the alphabet (which only gives 25 possible encodings), you could assign a *complete* re-mapping of the alphabet, for example

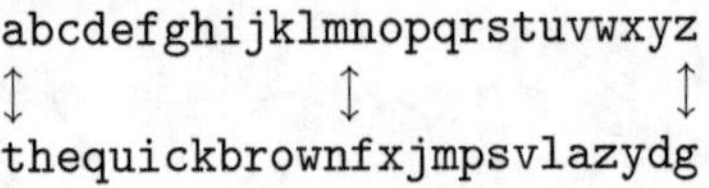

Note the use of a certain 'key phrase' (in this case the familiar *'The quick brown fox jumps over the lazy dog'*, which happens to contain all the letters of the alphabet at least once) to make it easier to remember the 'key' to the cipher. When using complete re-mappings (*transpositions* in cryptography jargon), the number of possible encodings is suddenly quite a bit more than 25 (you can use your `fac()` function to calculate how many, exactly).

This 'transposition cipher' was considered unbreakable for hundreds of years, until in the ninth century A.D. the Arabic scientist al-Kindī published a revolutionary approach. He made use of the fact that not every letter appears in a given piece of text equally often. For example, the letter 'a' appears in English text more often than the letter 'z'. By simply counting the occurrences of each letter in a large body of text in a given language, you can calculate quite accurate *frequency tables*. For example, it turns out that in English, the letter 'e' is the most common letter, occuring about twice as often as the letter 's', and that the letters 'q' and 'z' share the lowest popularity with each only one occurrence in every thousand letters or so.

By counting occurrences of all the letters in the encrypted message and comparing them with these frequency tables (along with some extra puzzling once you start to recognize certain words), it is often possible to break the code. The longer the message is, the more accurate the frequency tables of the code become, and the easier it is to decode the message.

In 1586, the French diplomat Blaise de Vigenère published a method immune to this frequency analysis. He shifted each letter in the message by a *variable* number of positions, as given by the current letter in a certain keyword (or key-phrase). For example, if the current letter in the keyword was an 'a', the letter would be be shifted by one position, if it was a 'b' by two positions, etc. For the next letter in the message, the next letter in the keyword was taken, wrapping around if the message was longer than the keyword (as is usually the case). This means that a particular letter in the original message can be encoded by several different letters in the encrypted text, depending on where in the text it appears and what the current position in the key-phrase is at that point. For example:

```
they are out to get you
open ses ame op ens esa
hwil svw pgx hd krl cgv
```

In this case, the key-phrase *'open sesame'* was used. You see that there are three e's in the original message, but they are encripted as an 'i', 's', and 'n', respectively. This method of encryption is called a *repeated key cipher* or *Vigenère cipher*, although Vigenère didn't invent it. He did invent an even stronger cipher now known as the *auto-key cipher*, in which the message *itself* is used in the key. This auto-key cipher was indeed unbreakable for 200 years until it was solved by Charles Babbage, which is worth mentioning here because Babbage is also considered the 'father' of the computer, having designed the 'difference engine' and 'analytical engine', entirely composed of mechanical parts.

We'll look at an implementation of a repeated-key cipher next. The drawback of variable shifting (or shifting with any other value than 13, in fact) is that the encryption and decryption algorithms have a sign change (if encryption adds, decryption needs to subtract). Since it would be nice to have a single filter for both encryption and decryption, we'll use the 'xor' operator instead, since `(m ^ k) ^ k = m`: xor'ing with the same key twice returns the original values.

```
#include <stdio.h>

int usage(const char *progname)
{
    fprintf(stderr, "Usage: %s <key>\n", progname);
    fprintf(stderr, "En/deciphers stdin to stdout using <key>\n");
    return 1;
}

int main(int argc, char *argv[])
{
    int c;
    int i = 0;

    if (argc != 2 || !strcmp(argv[1], "-h"))
        return usage(argv[0]);

    while ((c = getchar()) != EOF)
    {
        if (c >= ' ')
        {
            c = ((c - ' ') ^ (argv[1][i] - ' ')) + ' ';
            if (!argv[1][++i]) /* take next character in key */
                i = 0;         /* wrapping around if necessary */
        }
```

```
        putchar(c);
    }
    return 0;
}
```

This implementation first checks whether the character to be encrypted has a numerical value greater than or equal to that of the space character. This excludes the so-called *control codes* from being encrypted (such as the tab character, carriage return, and newline characters). Also, it adds the numerical value of the space character after xor'ing. This is because when the character to be encrypted and the current character in the key happen to be identical, the resulting encrypted value would be zero (`x ^ x = 0`), and we'd rather not have zeros embedded in the resulting string of characters (subsequent C programs reading the string could get confused by those because they think the string ends there).

In the code to take the next character in the key, notice how the key index `i` is *pre-incremented*, and the corresponding character is *then* checked to see if the end of the key string has been reached. The index is then reset to the first character in the key again.

Also, note that the resulting output of this implementation is not limited to ordinary letters, since it can result in values greater than the numerical value of the lower case 'z', which is the last 'ordinary' character in the ASCII encoding. If you print out the resulting encrypted text to the screen, you may see 'strange' characters such as accented or Greek letters[3].

8.5 Precedence

In §3.3, it was briefly mentioned that C follows normal precedence rules of calculus, so that `1 + 2 * 3` is evaluated as `1 + (2*3)`. By now, we have seen *many* more operators besides basic addition, subtraction, multiplication, and division. Each of the operators in C has a certain precedence, and the table below lists them all, hightest precedence to lowest:

```
()  []  ->  .
!  ~  ++  --  +  -  *  &  (type)  sizeof
*  /  %
+  -
<<  >>
<  <=  >  >=
==  !=
&
^
|
&&
||
?:
=  +=  -=  *=  /=  %=  &=  ^=  |=  <<=  >>=
,
```

[3]The drawback here is that this part of the character table is not fully standardized.

The +, -, and * at the top of the table are the *unary* versions of these operators. Using this table, you can see that for instance when masking bits, you should be careful:

```
return a&3 == 1; /* bit pattern of a ends in 01? */
```

because the bitwise 'and' operator & has a lower precedence than the equality operator ==, so the expression would be parsed as

```
a & (3 == 1)
```

which in turn becomes `a & 0` which is always zero. Also note the situation for 'multiple comparisons':

```
if (a == 3 && b == 2 || a == 2 && b == 3)
```

is parsed as

```
if (((a == 3) && (b == 2)) || ((a == 2) && (b == 3)))
```

The fact that && has a higher precedence than || is not generally realized, so it doesn't hurt to put some extra brackets in expressions like these.

Also, note that in the assignment-followed-by-comparison (see page 183), brackets are required around the assignment because '=' has a lower precedence than '=='.

8.6 Synopsis

C offers the basic 'binary logic' operators 'not', 'and', 'or', and 'exclusive or'. They are represented by ~, &, |, and ^, respectively. When applied to (integer) numbers, they operate in a 'bitwise' fashion, *i.e.*, the operators are applied to each bit of the number (in the case of the 'not' operator), or to each corresponding bit of both parameters (in the case of the three infix operators).

The three infix operators &, |, and ^ can be combined with assignment: `a |= 1` sets the least significant bit of a to 1 (regardless of whether it was 0 or 1 before the assignment), and `a &= 3` selects ('*masks*') the two least significant bits of a.

Compilers usually offer various *optimization levels*, so you can get a faster-running program (at the expense of a longer compilation time). Some optimization options are CPU specific.

Making your program run faster can often be achieved by avoiding calculations, for instance by 'remembering' results and simply looking them up next time they are needed instead of calculating them again. This works if the number of calculations is large compared to the number of possible input-output pairs.

Sometimes you can win a few percent of performance by keeping the hardware architecture in mind while implementing your programs. This is rarely worth the effort, as compilers are getting better and better at performing these kinds of optimizations.

8.7 Other Languages

Most languages have bitwise operators, and most use the same symbols for them as C does.

Java has an 'unsigned shift' operator >>> to mitigate the problem with shifting negative values. Since Java does not run on a 'real' machine but on a 'virtual' machine instead, the problem that not all (real) CPUs offer both 'flavors' of shifting does not apply.

Some languages are easier to optimize than others, and C is not necessarily the easiest for a compiler. Still, C compilers are usually among the most efficient for any given platform, because they have had the largest amount of effort poured into them.

For example, purely functional programming languages do not have the problem that the compiler does not know about possible 'side effects' of function calls. In C, the fact that a function can have side effects limits the compiler in its ability to move them outside of loops, for instance. In functional languages (such as ML or Clean), functions don't *have* any side effects. They can only return a value, or at the most modify their parameters. Of course, some functions *do* have side effects (because they print something to the screen, for instance), and this is sometimes solved by passing a 'world' parameter to them. A function which has side effects, after all, 'changes the world'.

The fact that languages such as Java or C# run in a 'virtual machine' at first glance seems to be detrimental to performance. After all, there is an extra intermediate step in executing the program. However, virtual machines have the advantage of being able to optimize *at runtime*, meaning they can focus on so-called *hot spots* of the program (pieces which are executed very often), and they can more easily optimize *across function calls* and even *across modules* (see chapter 10). For instance, it would be theoretically possible for a virtual machine to 'notice' that a certain 'expensive' function is called very often, and replace the function call with a LUT all by itself.

8.8 Questions and Exercises

8.1 Write functions to convert an integer number into its binary representation (*i.e.*, a string of ones and zeros) and *vice versa*. There are various ways of doing this, but you are obviously supposed to use the bitwise operators here.

8.2 How many *distinct* solutions does the puzzle represented on page 177 have?

8.3 ⋆ As mentioned in §8.3, the type of 3D puzzles presented here is a natural match for a recursive type of solution. Implement a program which generates only good configurations by starting with two pieces which match and trying to add new pieces to it one by one.

8.4 How much faster is your implementation from exercise 8.3 than the 'brute force' program from §8.2?

8.5 The first person to write programs for Babbage's machine was a certain lady, after whom the programming language Ada was named. Her last name (in all lower case) was used as the key in a Vigenère XOR cypher to encrypt the text of the next exercise. In case you can't find her name, it's Ybirynpr in rot13.

8.6 %)v<#4c&-!v7) 'e8'?6'a:*9o>$:$c$l,97>$ 1 6v2#3(,"(v=#3c3%(3+)3&e/6&-)3 37#($$!oc&#!17-56)-;?*"2b

Chapter 9
Simulations

What happens if a big asteroid hits earth? Judging from realistic simulations involving a sledge hammer and a common laboratory frog, we can assume it will be pretty bad.

Dave Barry

9.1 Virtual Experiments

In science, theory and experiment are two equally important facets in trying to understand the world around us. On the one hand, experiments are devised to test certain hypotheses. On the other hand, sometimes experiments yield results which call for a new theory.

In many cases, experiments can be performed 'in real life'. Ideally, one can perform a series of trials varying only a single parameter, and establish a correlation between this parameter and the observed behavior. An example would be dropping a series of balls from a certain height, all with the same dimensions but having a different mass, to see whether its mass has any influence on the time it takes for a ball to hit the ground. Another example would be determining the solubility of some solid in a liquid as a function of the temperature, or letting a series of seedlings grow in varying lighting conditions to observe the effect of light on plant growth.

In many other cases, experiments are not so easy to do. Imagine an astrophysicist wanting to test a theory on the behavior of colliding galaxies. One can't simply 'set up' such an experiment, and if there doesn't happen to be any astronomical data available for such an event, the scientist is more or less 'out of luck'.

When this astrophysicist is willing to make some assumptions though, such as 'galaxies are so enormous and so sparse that their individual stars can be thought of as having their mass concentrated in a single point', and 'stars adhere to Newtonian gravitation mechanics', he could sit down with a stack of paper and *calculate* what would happen. For 'simple' systems of interacting particles, this is quite feasible. However, even when there are only a few particles involved, this rapidly becomes undoable. Let alone when we are dealing with hundreds of thousands of particles, each interacting with all the others.

Here, computer simulations can help. In this chapter, we'll look at a few 'computer-simulated experiments'. First, a word of caution. When performing experiments 'in the real world', if you forget a certain important aspect in your hypothesis, your experiments will most likely prove it wrong (unless you forget *two* important aspects which happen to cancel each other out). In a computer simulation however, if your program is wrong, your results will be wrong. Sometimes this can be on purpose: 'Suppose we eliminate the effects of friction, what would happen?' But sometimes this is simply an error, rendering the entire experiment meaningless. It is therefore often advisable to start out with a setup for which the outcome is known, so you can at least check whether your simulation is in accordance with reality for that particular case before heading off in unexplored territory.

There won't be many new language features treated in this chapter, but with the material covered so far you can already write quite a few useful scientific programs.

9.2 A Pendulum

In this section, we will examine a *pendulum:* a mass suspended on a wire, swinging back and forth. This is a nice first example, because it is straightforward to derive an analytical solution to this problem in the case of 'relatively small' amplitudes. This makes it possible to compare the simulation results with the 'known' solution before extending the simulation to areas where the analytical solution is considerably more difficult.

9.2.1 Analytical Derivation

Consider a mass m suspended on an (inelastic, mass-less) wire of length ℓ. The earth pulls on this mass m with a gravitational force $F_g = mg$, and when it is in 'resting position', this force is exactly cancelled by the pulling force in the wire F_w. When the mass is moved out of its center position by a distance x, the gravitational force and the pulling force in the wire no longer cancel, and there is a remaining force F_r pulling the mass back to the center position. This is shown in the figures below.

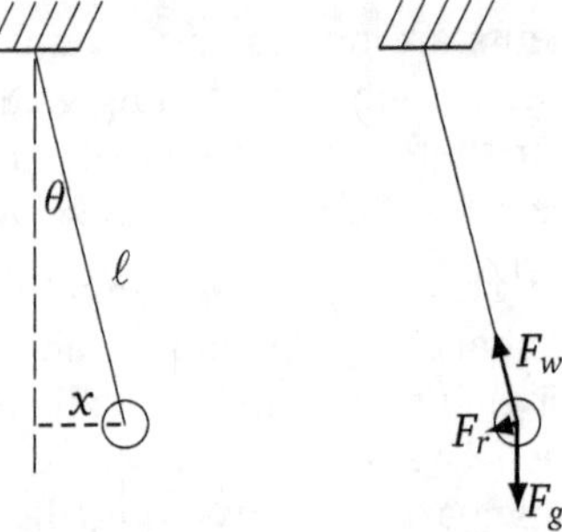

Simple trigonometry shows that $F_r = -F_g \sin\theta = -F_g x/\ell$. The mass will be accelerated by $a = F_r/m$, and since $F_g = mg$, the end result is that $a = -gx/\ell$.

Now, for small θ (*i.e.*, when the pendulum is moved only a little bit from its resting position before it is let go), we can ignore the fact that a and x are not entirely parallel, and simply state that

$$a = \frac{\mathrm{d}^2}{\mathrm{d}t^2}x(t) = -\frac{g}{\ell}x(t),$$

which is a familiar differential equation: Some function $x(t)$, twice differentiated, yields the same function again but with a minus sign and some factor. We know the solution to such a differential equation: the sine (or cosine) function. It is more natural in this case to take the cosine, since at $t = 0$ we are releasing the pendulum from a nonzero starting position x_0:

$$x(t) = x_0 \cos\left(\sqrt{\frac{g}{\ell}}\, t\right).$$

It is usually surprising to physics students when they first find out that the mass of the pendulum has no influence on its frequency (which is $\frac{1}{2\pi}\sqrt{g/\ell}$), nor does its initial position x_0.

9.2.2 Numerical Simulation

We will characterize the pendulum by its position x and its velocity v (still pretending that the problem is essential one-dimensional). To visualize the position of the pendulum, we will use some crude graphics:

```
void print_pos(double x)
{
    int i;

    if (x > 39 || x < -39)
        return; /* position doesn't fit on screen */

    for (i = 0; i < 40 + x; ++i)
        printf(" ");

    printf("*");
}
```

Assuming a screen width of 80 columns, this routine prints an asterisk at the position of the pendulum (so we are limited to positions ranging from −39 to 39, putting the 'resting position' in the middle of a line). Note that the function does not print a 'newline' character (`\n`) after the asterisk, which you might have expected – we'll see why below.

The 'body' of the simulation consists of an endless loop which uses the function above to print out the current position of the pendulum and then updates its position and velocity:

```
int main(void)
{
    double length = 200;
    double g = 9.8;
    double x = 38;    /* initial position */
    double v = 0;

    while (1)    /* endless loop */
    {
        double Fr = -g*x/length;

        print_pos(x);
        v += Fr;
        x += v;

        getchar();
    }

    return 0;
}
```

When you run this program, a new value is calculated (and displayed) each time you press the enter key. You'll have to exit the program by pressing Control-C or Control-Break (depending on whether you are on a UNIX system or a Windows system). Since pressing the enter key normally also causes a newline to be displayed on screen, we didn't have to do this ourselves in in the `print_pos()` function.

The output on screen after a few runs should look somewhat like this:

```
                                                                *
                                                               *
                                                            *
                                                        *
                                                  *
                                            *
                                    *
                              *
                       *
                *
          *
      *
   *
 *
*
  *
    *
        *
              *
                    *
                           *
                                   *
                                         *
                                                *
                                                      *
                                                          *
```

This looks quite nicely like the expected cosine wave.

9.2.3 Extending the Simulation

In the current model, the pendulum will keep swinging forever. In reality, this is not so, and that is mainly due to friction of the mass as it passes through the air surrounding it.[1] Therefore, let's extend our model of the pendulum to include friction.

The mass at the end of the pendulum experiences a frictional force which is proportional to its velocity squared:

$$F_f = \frac{1}{2}C_f A \rho v^2$$

where C_f is a constant depending on the shape of the object (the factor car designers try to get as low as possible), A is its area perpendicular to its direction of movement, and ρ the density of the medium the element is passing through. For our purposes, we can simplify this by combining all the constants:

$$F_f = Cv^2$$

where C becomes another parameter of our model.

Now, we modify the expression calculating the acceleration by subtracting the frictional force from the force exerted by the neighboring elements, and using this new resulting force instead:

```
double Fr = -g*x/length;
double Ff = C*fabs(v)*v;
double Ft = Fr - Ff;
v += Ft;
```

The function `fabs()` returns the absolute value of its `double` argument. It is part of the standard C library and can be used by including `math.h`. We've seen how such a function could be implemented in §4.1.

We use this function because otherwise the resulting friction force would always be positive, in which case negative velocities (*i.e.*, the pendulum moving 'to the left' in our case) would experience an *extra* acceleration due to friction, instead of a deceleration.

Setting C to about 0.02 gives a nice 'dampening' effect.

Before moving on to the next section, it is probably a good idea to take a look at the first few exercises at the end of this chapter. Especially question 9.4 is important, since it deals with an important limitation of numerical simulations.

9.3 Random Values

In §3.6, an example was given of a function with a return value but without any parameters. The example there was a `random()` function, and such a function exists indeed. The function `rand()` returns a (pseudo) random integer value between 0

[1]This is not the *only* reason; a pendulum in vacuum still doesn't keep swinging *forever* as it loses energy due to internal friction as well. We will ignore this effect because it is much smaller than the air friction.

and RAND_MAX (the definition of RAND_MAX is in stdlib.h). Of course, it is impossible to 'calculate' a random value, so this is *pseudo* random: making clever use of the characteristics of the way numbers are handled in computers, including their overflow behavior, the result of successive calls to rand() are a series of numbers which are *in principle* deterministic, but which *appear* to be random. For *real* random values, there should be a hardware device connected to the computer which delivers the values. You can even order CD-ROMs containing millions of random values, which were taken from physical random generators. For many purposes, however, pseudo-random values suffice.

Often, it is easier to deal with floating-point random values in the interval $[0, 1)$; for this, you can use the macro[2]

```
#define frand() (rand()/((double)RAND_MAX + 1))
```

We will use this function to simulate a 'marble board': Take a wooden board and hammer in nails in the shape of a triangle, with one nail at the top, two on the second row, three on the third, etc. (see the figure on the next page). Then, drop a series of marbles from above onto the topmost nail. Each marble will bounce on the topmost nail, either fall to the left or the right of it, then bounce on one of the nails on the second row, bounce to its left or its right again, etc. When it finally reaches the bottom of the board, each marble has made left or right bounces as many times as there are rows in the board.

Given enough marbles and a big enough board, the resulting positions at which the marbles drop off the board at the bottom form a *Gaussian distribution:* If the nails are placed at the exact right positions, the chances of bouncing either left or right are equal, and the chances of ending up near the middle of the board, after an approximately equal number of 'left bounces' and 'right bounces', are much greater than the chances of making only left bounces and ending up at the far left of the board, or only right bounces and ending up at the far right. This is used in several games of chance, of which the 'pachinko' variety is very popular in Japan.

The figure on the facing page shows a 'marble board' with 36 nails, and one possible trajectory of a marble exiting the board in the middle.

The way we will simulate this is by virtually dropping marbles, making a series of 'left-right' decisions using the frand() macro we have just seen, and noting the current position as we go. We will use a similar trick to draw the resulting distribution on the screen as we used in the pendulum simulation of the previous section. We will place a virtual row of 'bins' below the virtual marble board in which each dropped marble ends up, by increasing the value in in an int array corresponding to the 'bin' the marble ends up in, and we will use a (large enough) number of experiments to build up the probability distribution.

---

[2]In the original K&R book, this macro is defined slightly different, but the version presented here works also on systems where RAND_MAX is equal to INT_MAX.

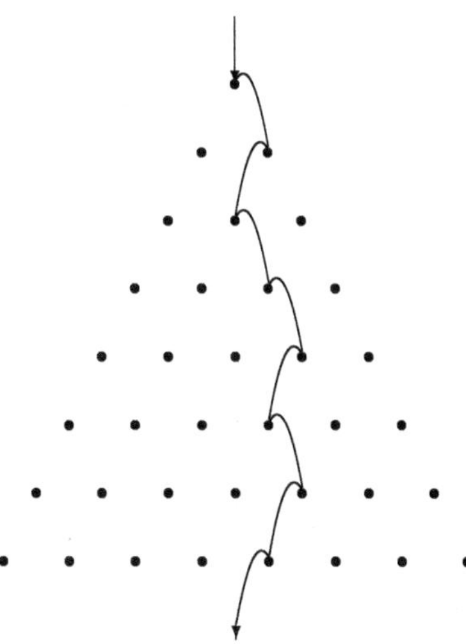

First, let's set up the distribution to be all-zero using a fixed number of 'bins'. This number is a measure of the size of the marble board, and it is a parameter of our simulation. Since we will use a similar trick to display the resulting distribution as we did in the pendulum example in the previous section, we'll pick a number which nicely fills one screen-full.

```
#include <stdio.h>
#include <stdlib.h>

#define BINS 24
#define N_EXPERIMENTS 400

#define frand() (rand()/((double)RAND_MAX + 1))

void draw_distribution(int distribution[], int size)
{
    int i, j;
    for (i = 0; i < size; i++)
    {
        for (j = 0; j < distribution[i]; j++)
            printf("*");
        printf("\n");
    }
}

int main(void)
{
    int distribution[BINS];
    int i;

    /* clear the initial distribution */
    for (i = 0; i < BINS; i++)
        distribution[i] = 0;
```

```
    /* perform the simulation a number of times */
    for (i = 0; i < N_EXPERIMENTS; i++)
    {
        int j = BINS;     /* j tracks the current position */
        int k;

        /* "walk" the marble board */
        for (k = 0; k < BINS; k++)
        {
            if (frand() < 0.5)     /* flip a virtual coin */
                j++;     /* take a right */
            else
                j--;     /* take a left */
        }

        /* note where the marble exited */
        distribution[j/2]++;
    }

    draw_distribution(distribution, BINS);

    return 0;
}
```

The chances of `frand()` returning less than 0.5 are 50%, so we are giving equal probabilities to either a left-bounce or a right-bounce. Note that we start out with `j = BINS` and we increase the distribution at `j/2` at the end of each simulation, instead of starting out at `BINS/2`, the 'real' center position, and increasing the distribution at bin `j`. This is because `j` is either increased or decreased at each bounce and never stays unchanged, meaning that depending on the number of bins we use, either only the even bins or only the odd bins get filled, resulting in a distribution with 'holes' in it.

If you run this program, something like the following distribution should appear on your screen:

```
*
***
*******
******************
******************************
************************************************
************************************************************
********************************************************
********************************************************
************************************************
****************************************
*****************
************
****
```

This approaches the familiar bell-shaped Gaussian distribution (tipped on its side).

However, if you run the program again and again, you may notice something interesting: You get the same curve each time! How is this possible, given that the decisions are supposed to be random?

The reason is that the `rand()` function returns *pseudo*-random numbers, as was mentioned at the start of this section. The algorithm behind `rand()` works by 'remembering' the previous value returned, and in each successive call performing a calculation using this number which causes an overflow, so that the resulting new number appears to be unrelated to the previous one. However, the calculations are the same each time, so if a series starts out with a certain value, called the *seed* of the random generator, the resulting series of pseudo-random numbers will be identical. At each program start, the seed is reset to the same value, hence you'll get the same random series each time you run the program.

Sometimes this is desirable, because this way you can 'replay' an experiment; but in this case, we'd rather have new results each time we try the simulation. We can arrange this by setting the seed ourselves (using the `srand()` function). The problem is that we need to figure out a different seed each time we run the program, so we are facing a 'chicken-and-egg' problem! There is no use calculating a random value for this seed, because that random value will be the same every time we run the program...

One thing which *does* change from run to run is the current *time*. Most computer systems have an internal clock, and C offers functions to read out its value, often in the form of a single integer number representing the number of seconds passed since the 'beginning of time', for example January 1st, 1970 – an arbitrary choice, and not even the same one on every system, but that doesn't matter as long as it is a different value each time we run the program. If we `#include <time.h>`, we have the function

```
time_t time(time_t *tp);
```

It returns the current time[3]. If `tp` is non-`NULL`, the current time is also assigned to `*tp`. By using this value as the seed of our random generator, we can get a different series each time we run the program. Simply adding

```
srand(time(NULL));
```

to the beginning of our simulation makes sure we see different distributions at each program run, unless we run the program quicker than the granularity of `time_t`. On most systems this is one second.

Incidentally, the `time.h` header file also makes available a whole range of other time-related functions, including functionality to convert a `time_t` value to a human-readable 'calendar' string, and for dealing with time zones and daylight saving. More information will be given in §12.3.4.

Finally, it is worth noting that many systems offer a random generator with significantly better statistical properties than the simple `frand()` macro we were using here. For some situations, this can be important. For example, for the simulation of

[3]Officially, it may return -1 if there is no system clock available on this particular system, but unless you are on a really exotic or ancient system, this is not an issue.

this section it wouldn't be too much of a problem if the random generator had the property that every, say, tenth 'coin flip' would always be the opposite result of the one immediately preceding it, as long as the chances for 'heads' or 'tails' would be approximately equal for the rest of the coin flips. However, you wouldn't use such a random generator to run an online casino, because such a 'pattern' would soon be found out and taken advantage of.

9.4 Percolation

In this section, we will use simulations to examine a class of phenomena known as *percolation*. In the most elementary sense, the problem consists of finding a set of connected locations (*i.e.*, a pathway) through a medium. Let us start with a 2D example in which we have an equal-spaced grid. On each grid location, there can be a 'pass-through' or a 'block'. The distribution of these is determined by a single parameter: the chance of there being a 'pass-through' at any given location. In the figure below, a small 8×8 grid is shown, where open circles denote the 'pass-through' locations, and closed ones denote the 'blocks':

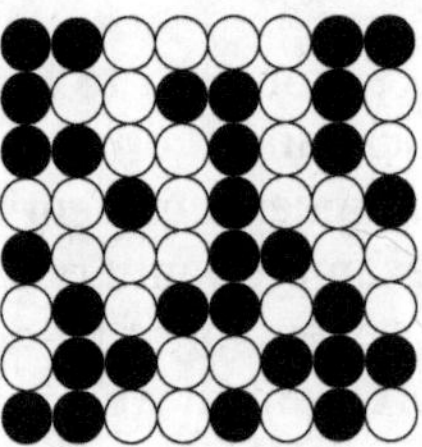

This particular grid has 31 blocks and 33 pass-throughs, so apparently the pass-throughs were distributed with approximately 50% probability. Below is the same grid, but with a pathway shown which connects the left side of the grid to the right side:

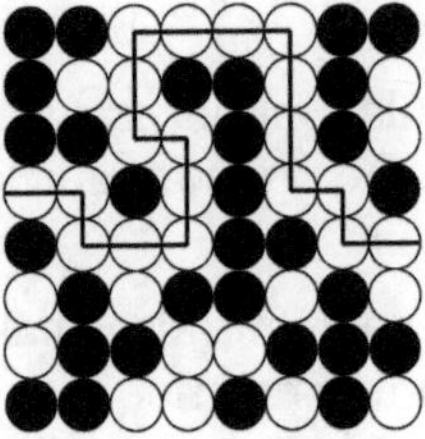

The goal of our simulation is to find pathways like these automatically, and figure out what the chances are of such pathways existing as a function of the distribution of 'pass-throughs' in the grid. We know by common sense that a pass-through distribution probability of 0 yields a zero chance of finding any pathways, and a probability of 1 gives a 'completely open field'. What the curve in between looks like is interesting, and this is the subject of this section.

This may seem like frivolous 'maze-solving', but there are interesting real-world applications in which similar situations apply. For example, if the 'pass-throughs' represent conducting particles and the 'blocks' represent insulating particles, then simulations like these help figure out what the chances are that a certain mixture of conducting and non-conducting particles is electrically conducting as a whole. Another example is to estimate the probability of a certain thickness of a porous, foam-like material with a certain density being airtight.

To do the maze-solving part, we will use the recursion concept treated in §4.9 along with a new 'trick' called *backtracking*.

First, we will set up the grid. We like to print out the resulting solution to the screen, so we take a grid size which nicely matches the screen. We fill a two-dimensional `int` array (see §5.4). At each grid position, a zero means a block (or a non-conducting cell, or a piece of solid material in a foam), and a 1 represents a 'pass-through' (or a conducting particle, or an air bubble). We will first look at a program generating and displaying a single grid:

```
#include <stdio.h>
#include <stdlib.h>
#include <time.h>

#define WIDTH 79
#define HEIGHT 24

#define frand() (rand()/((double)RAND_MAX + 1))

void draw_grid(int grid[][WIDTH])
{
    int i, j;
    for (i = 0; i < HEIGHT; i++)
    {
        for (j = 0; j < WIDTH; j++)
        {
            if (grid[i][j] == 0)
                printf(" ");
            else
                printf(".");
        }
        printf("\n");
    }
}

int main(int argc, char *argv[])
{
    int grid[HEIGHT][WIDTH];
    int i, j;
    double chance = 0.5;
    if (argc == 2)
        chance = atof(argv[1]);
```

```
    srand(time(NULL));
    for (i = 0; i < HEIGHT; i++)
    {
        for (j = 0; j < WIDTH; j++)
        {
            if (frand() < chance)
                grid[i][j] = 1;
            else
                grid[i][j] = 0;
        }
    }
    draw_grid(grid);
    return 0;
}
```

If you run this program a few times with different probability values, you will get a feeling for the 'density' of such grids.

Incidentally, the loop filling the grid could have been written a lot more compact (and cryptic) as

```
for (i = 0; i < HEIGHT; i++)
    for (j = 0; j < WIDTH; j++)
        grid[i][j] = (frand() < chance);
```

See §4.3 about boolean expressions if this doesn't make sense straight away.

Now for the 'meat' of our simulation: automatically finding out whether there is any pathway from the left side of the grid to the right side.

Looking at this problem with recursion in mind, the notion of a 'pathway to the right side' can be formulated a little differently: *A certain grid point is part of a pathway to the right side if at least one of its neighboring grid points is part of a pathway.* In other words: *You can reach the other side of the grid from a certain point if you can reach the other side of the grid from one of its neighbors.*

Therefore, we will simply walk along all points on the left side of the grid, and determine whether any of these fulfills the descriptions above. If so, we can stop: we needn't find *all* pathways nor the shortest.

Let us construct the recursive function to do this 'pathway check'. First, let's define its return value: It should be zero if it's not part of a pathway, and one if it is:

```
int percolate(int grid[][WIDTH], int i, int j)
{
    if (i < 0 || j < 0 || i >= HEIGHT)
        return 0;    /* invalid position: fell off the grid */
    if (i >= WIDTH)
        return 1;    /* we have reached the right side! */

    if (grid[i][j] == 0)
        return 0;    /* we ran into a block */
```

```
        /* check whether our neighbors are part of a pathway    */
        if (percolate(grid, i, j + 1)     /* right neighbor?     */
         || percolate(grid, i, j - 1)     /* or the left one?    */
         || percolate(grid, i + 1, j)     /* or below us?        */
         || percolate(grid, i - 1, j))    /* or above?           */
            return 1;   /* If so, we're part of the pathway too  */

        /* otherwise, this is a dead end */
        return 0;
    }
```

It may be strange to see four recursive calls of the same function, but you are free to make as many as you like. It can be mind-boggling if you try to perform this 'in your head', but the compiler does a good job of it so you don't have to.

The bottom part of the function performs the test we described above: If we aren't on a block, and one of our neighboring grid points is part of a pathway, then we are part of a pathway too. The top part of the function makes sure we don't run off the grid, and takes care of stopping the recursion when we reach the other side of the grid. Be sure to let this function 'sink in' for a while, perhaps using the example grid presented earlier in this section, and see whether you can develop a good feeling for how this algorithm works.

It may be helpful to keep the following analogy in mind. Suppose you are dropped at the entrance of a maze, and your task is to say whether this maze has an exit. Further suppose that you have the capability of splitting off identical clones of yourself. One way to solve your task is to split off a clone of yourself and send him into the maze, with simple instructions:

1. At each junction in the maze, pick one corridor for yourself and send clones of yourself into the others, giving each of these clones this exact list of instructions

2. report back at the previous junction when you either run into a dead end or find the exit

3. when you arrive back at the previous junction, wait for all your clones to arrive there too and 're-join' with them into a single entity again, remembering whether you or any of your clones has found the exit, and finally

4. report back at the previous junction.

When a clone returns to his spawning point, he can either report that "one of the paths I just came from reaches an exit" or "none of the paths I just came from reaches an exit". Finally, the first clone you split off comes walking out of the maze and reports to you whether *any* of his clones has found an exit.

With the help of the analogy, perhaps you can also find the fatal flaw in this algorithm before reading the next paragraph.

This algorithm will quickly end up in infinite recursion. Why? Because a clone cannot determine whether a certain maze junction has been visited before. One of the clones is bound to reach a junction already visited, split off new clones, and send them off following 'older clones'. In fact, one of them will run into the very corridor his 'parent' just came from! Clearly, recursion alone isn't enough to solve this task.

Enter *backtracking*. This is the computational equivalent of the fairy tale of Little Thumbkin, leaving a trail of bread crumbs on his way through the forest. What we have to do is *mark* the grid points we've already visited. Below is a modified version of the `percolate()` function which does just that.

```
int percolate(int grid[][WIDTH], int i, int j)
{
    if (i < 0 || j < 0 || i >= HEIGHT)
        return 0;
    if (j >= WIDTH)
        return 1;

    if (grid[i][j] == 0)
        return 0;

    /* have we visited this point before? */
    if (grid[i][j] > 1)
        return 0;   /* in that case, don't bother continuing */

    /* Mark this grid as visited */
    grid[i][j] = 2;

    if (percolate(grid, i, j + 1)
     || percolate(grid, i, j - 1)
     || percolate(grid, i + 1, j)
     || percolate(grid, i - 1, j))
    {
        grid[i][j] = 3;     /* mark as visited AND part of pathway */
        return 1;
    }

    return 0;
}
```

We need to modify the `draw_grid()` function a bit to deal with these 'marked' grid points:

```
void draw_grid(int grid[][WIDTH])
{
    int i, j;
    for (i = 0; i < HEIGHT; i++)
    {
        for (j = 0; j < WIDTH; j++)
        {
            if (grid[i][j] == 1 || grid[i][j] == 2)
                printf(".");
```

```
                else if (grid[i][j] == 3)
                    printf("*");
                else
                    printf(" ");
            }
            printf("\n");
        }
    }
```

Below is a typical output of the program with a pass-through density of 0.7:

```
 . .... .......... .... .... ... . ... ...... . .... ....... .. .  ..... . .....
.... ... .. .......    .   ... ......... .... . ..... .. .  ....  ..     ... ....
......  . .... ....    . .  ..... .. ............ .. ...........  . ... . ... ..
  . . . ......... ... .... ..  . ......... ....... ..... .  .....  . . .. ....
   .  ..     . . ... ..  .. . ... .. .  .  .  . ..  . . . ..   ...  .......... .
********* . ...   ..... .  .. ... .... ............. . .. .. .. ... ..   ...  ...
. ... ..**. . .  ...... ......   .... ...........  .. . . .. ..... . .  ...  ..
........ *   ...... .  ..  . ....     ..  ....... . ******* ...  ........... .....
   ......** .   . . . .. .. ......  ... ... ..  ..*** . ****  ......... .. .  ..
  .. .  ..******  ... ... ............  ...   .. ***.  .. ************* . ....
.. ...  .. . ..**  ..  ... .  ... . .... ... .... ****   .. ..  ... .  ** .*****
.....  .**** ****. .... ..  ....... ......... ...  ***  ....  ***********  *   .
. . ..  ** *** .. .  ..... . ............. .... . .** . ..   .**   ........*.. .
..  .... *.   .. .  .... .. ....... ... .  . . ...*** . .. . .** .. .  ..***  .
........ * .   .. . .. . . .. . ...... .........***  ......   *. .. .. . *** . .
 . . ....****** .. ........ .. .    . .... ******  ..  . ... ** .... *******.. .
... ........   *  ..  . . ....********** . ** ... .. .. .  .  .****.   * .. .. ..
... ...... ...***. .... .. . **** ....***** ... . .........    ** . **.  . ..
 ....... .. ... *  ... .....*****   ......  ...  . .. ... . ****. ..*   .     ..
 ......... . ...**  .......** ****  ............. ... . ... .*** ..  ***.. .....
 ..... ...... ...**  ...... **** **. ..   ... . ..  ... ..  ...**  ... ** .. .. .
. .. ....  ....  .**********  **  ......... ...... .  .   ..*********. . . .
............. ........   ...****  .... . ... .  ....  ........ ... . ..  ......
 .... .  ....... .. ....  ..... . ....... .. .......  .. ....  .   ... .  .......
 ......   .. ... .. ... .........  .... . . ...  ... .  . . . .. .. .. ..... ..
```

Note that the code above simply finds the *first* pathway, not necessarily the shortest. For our current experiments, we aren't particularly interested in the length of the pathway; just in whether there *is* one.

There are several questions and exercises at the end of this chapter, starting with exercise 9.7, in which the simulation of this section is used to examine some interesting properties of these grids.

As a final note: we are using an `int` array for the grid, and this is a bit 'overkill' since we only use four different values for each grid position (out of the four *billion* values available for an `int`). For the grid size we've been considering, this is not an issue. However, on many systems the amount of memory available on the stack is limited to a much smaller size than heap memory (memory available through `malloc()`, see page 110). If you are running the simulation with larger grids, or extend it to examine three-dimensional grids, you may get a compiler error that the `grid` array is getting too big for the stack. In that case, you could modify it to use `chars` instead of `ints`, which will buy you a factor of four. Ultimately, you will need to modify the program to work with dynamically allocated memory (see §5.7).

9.5 Synopsis

This chapter presented several 'real-world' examples of simulations and computer experiments. In all cases, it is important to find a mapping of a real-world situation to a model which is suitable for computer simulations, *i.e.*, a *discrete* model. It is important to realize that this discretization step has some implications to the reliability of the simulation, and there is often a trade-off between speed and accuracy. For an accurate simulation, you'd rather take very small discretization steps, requiring many iterations of the algorithm, and which may get 'stuck' due to 'underflow': The differences in each step are so small that they get lost in the limited precision of the computer. For fast results you would use large discretization steps, with the risk of 'overshoot' where the simulation may 'explode'.

In many simulations, *random behavior* is required. Since 'generating random numbers on a computer' is a contradiction in terms, you normally have to settle for *pseudo-random* behavior. The computer can generate a series of numbers which *appear* to be random; the quality of the random generator is measured by how easy it is to spot certain 'patterns' in the sequence of numbers.

A certain class of problems can be tackled by using 'maze-solving' algorithms. Here, an important extension to *recursion* is *backtracking*, in which the recursive search for solutions is augmented by marking which solutions have already been examined, and 'reversing the steps' once it turns out that a certain path doesn't lead to a solution.

A simulation using only 'first principles', is called an *ab initio* experiment. It can be dangerous to interpret the results of such simulations because you must be sure you haven't left out any important effects. In a 'real' experiment, you usually assume that the measurements are right – if they don't stroke with your hypothesis, it is quite likely that the hypothesis is incomplete. In a computer simulation, there may be other effects to deal with, like limited precision, overflow or underflow, or even simple miscalculations.

9.6 Other Languages

C was not designed with scientific simulations in mind, but the language does lend itself quite nicely to such experiments. For some experiments, an object-oriented language (see §6.5) might be more suitable, because you can directly map concepts of the real-world situation to classes of objects in the language.

Most languages offer sufficient mathematics support for most experiments, and most support recursion (and thus backtracking).

Traditionally, most 'scientific' programs are written in FORTRAN (which is older than C). In fact, its name comes from FORmula TRANslator, and it was specifically designed so that scientists could write programs which looked more 'mathematical' instead of using the machine language of a particular machine (see §1.2.1).

Because FORTRAN has been the 'language of choice' for scientists for several decades, there exists an enormous amount of 'mature' (*i.e.*, highly optimized and reliable) code in so-called *libraries* (see chapters 10 and 12), but it is often possible to call functions from these libraries from your C programs.

9.7 Questions and Exercises

9.1 In the pendulum simulation of §9.2, determine the period of the pendulum, either by simply counting the number of times you need to press the enter key before the pendulum reaches its original position again, or by modifying the program to keep track of the maximum position. How does it compare to the theoretical value? Also modify the length of the wire and see whether the resulting frequencies correspond to the theory.

9.2 Determine the effect of the 'friction term' on the frequency of the pendulum.

9.3 Varying the friction constant C in the pendulum simulation determines how quickly the pendulum is at rest again. When C is very low, the pendulum will not experience much friction and keep swinging for a long time; When C is above a certain value, the pendulum will not even swing at all (*i.e.*, it slowly approaches its resting position and never swings to the other side). This would correspond to a pendulum placed in a container full of syrup or some other viscous liquid. When C has the smallest value for which the pendulum *just* doesn't swing anymore, the system is said to be *critically dampened*. Determine this value of C.

9.4 The constant g used in the pendulum simulation is only valid here on Earth. On different planets, there would be a different g. Change the value of g to be ten times as big and see what influence that has on the frequency of the pendulum. Then, change it to be a hundred times as big (to simulate, say, a pendulum near a heavy star). What happens? Can you explain? How could you modify the code to handle extreme gravity like this?

9.5 ⋆ Making the length of the pendulum comparable to the initial position x_0 will render the derivation invalid, but the code of the simulation does not take this into account. Modify the simulation so that it takes vertical acceleration into account as well (*i.e.*, distinguish between v_x and v_y) and keep track of the vertical position of the mass. You will need to take centripetal acceleration into account ($a_c = v^2/r$), or the pendulum will keep 'falling down'. You may notice that it is very hard to get good results with your simulation, because the errors made in each step keep adding up, rendering the simulation useless.

9.6 In the marble board simulation of §9.3, modify the left-right decision criterion to give different chances for left and right (*i.e.*, use a 'loaded die') and observe how this changes the resulting distribution.

9.7 The first thing to examine with the 'percolation grids' of §9.4 is the probability to find a pathway as a function of the probability used to distribute 'pass-throughs' and 'blocks'. Examine all such distribution probabilities from 0 to 1 in increments of 0.05, perform several simulations at each probability, and plot the probability of finding a pathway. You will quickly find out that there is a 'critical density' around which the pathway probability rises steeply. What is this 'critical density' for the grid size used in the examples?

9.8 Obviously, the grid size is an important factor too. Examine the effect of grid dimensions on the 'critical density'.

9.9 In the clone analogy, one could argue that it would be much easier to simply leave one clone at each junction, who 'remembers' where he came from, and only send off his clones in the *other* directions. Modify the program to be more like this. Can you get it simpler this way?

9.10 ⋆ Modify the pathway-finding algorithm to find the *shortest* pathway through a given grid. Given this modified algorithm, plot the *average* pathway length against the pass-through probability, not counting simulations which couldn't find a pathway at all. Hint: the `percolate()` function as it is now only returns *whether* there was a pathway in this direction. Can you modify it to return the number of steps needed to reach it? At each grid point, compare the return values of `percolate()` and let the shortest non-zero value win. One is added (for the current grid point) and the result returned again.

Chapter 10
Complex Projects

A complex system that works is invariably found to have evolved from a simple system that works.

John Gaule

10.1 Multi-File Programs

The programs we have seen so far are all relatively short. They contain at most half a dozen functions, and the total source code fits on a few pages or screens.

However, C is certainly not limited to short, simple programs. In fact, programs written in C routinely have hundreds of thousands (or even millions) of lines of code. Of course, typing all this code in one single file would be very impractical, and compiling this huge file every time you make a change somewhere in it would be very time-consuming. In this chapter, we will revisit some of the materials from chapters 2 and 3 and see how they can be amended to deal with large, complex programs.

In C, you can split up the source code of a program over several files, which can be compiled separately. Recall the figure on page 34. There, the compiler translates *one* source file into *one* object file, but the linker can link *several* object files (hence its name) into an executable, as in the figure below. Here, the source files `a.c` and `b.c` are compiled and linked into an executable `foo.exe` (file name extensions are as on Windows):

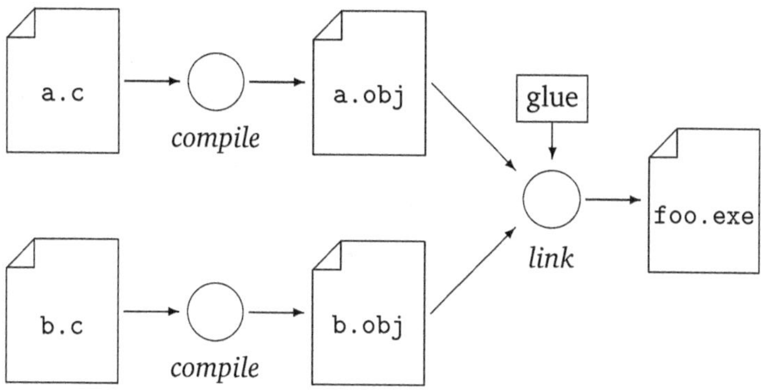

In fact, the 'glue' is usually just another object file.

Let us first get the technicalities of building multi-file programs out of the way. Consider a trivial variation of the 'Hello, world' program:

```
#include <stdio.h>

void hello(void)
{
    printf("Hello, world!\n");
}

int main(void)
{
    hello();
    return 0;
}
```

To get a handle on the sheer complexity[1] of this program, we will break it up in two source files. Let's put the `hello()` function in a file `greet.c` (including the `#include <stdio.h>`), and the `main()` function in a file `main.c`. Then, let us try to compile these two source files. If we try

```
cc -o hello main.c
```

we'll get an error somewhat like the following:

```
/tmp/cctNuXbJ.o: In function 'main':
/tmp/cctNuXbJ.o(.text+0x11): undefined reference to 'hello'
collect2: ld returned 1 exit status
```

On a Windows system using Visual C++, we would have tried

```
cl /Fehello.exe main.c
```

(the `/Fe`*filename* switch determines the name of the resulting executable) and there we would have gotten an error like

```
main.obj : error LNK2019: unresolved external symbol _hello
referenced in function _main
hello.exe : fatal error LNK1120: 1 unresolved external
```

The actual wording will vary with the system, but the main message here is that the linker was looking for a function called `hello`, which was called from `main`, and couldn't find one. (Of course it can't, because we moved it to `greet.c`.) Therefore, it can't link the program into a proper executable, and it gives up with an error message. If you are compiling with maximum warning level (you *are* normally doing that, right?), with something like

```
cc -Wall -o hello main.c
```

[1] That was a joke

we would have gotten a prelude to this error:

```
main.c: In function 'main':
main.c:3: warning: implicit declaration of function 'hello'
/tmp/ccnWYI5y.o: In function 'main':
/tmp/ccnWYI5y.o(.text+0x11): undefined reference to 'hello'
collect2: ld returned 1 exit status
```

We will talk about this warning in more detail later on, but lets us first focus on getting rid of the errors.

If we try to compile `greet.c`, we will get something like this:

```
/boot/develop/lib/x86/start_dyn.o: In function '_start':
/boot/develop/lib/x86/start_dyn.o(.text+0x5d): undefined reference
to 'main'
collect2: ld returned 1 exit status
```

Again, the error you get on your system can be quite different, but it should be clear that the linker can't find the `main()` function, and thus can't link this program into a proper executable either. (The `_start` function is part of the 'glue' on this particular system, which happens to be in an object file called `start_dyn.o`.)

Somehow, we have to compile *both* of these source files into a single executable. On most systems, the solution is quite simple: You can specify multiple source files in a single command line.

```
cc -Wall -o hello greet.c main.c
```

Or, with Visual C++:

```
cl /W4 /Fehello.exe greet.c main.c
```

This single command compiles both `greet.c` and `main.c` into object code, subsequently links those object files together, throws in some glue, and produces the resulting `hello` executable.

It doesn't get rid of the warning mentioned before. That is because when the compiler translates a source file, it starts out with a 'clean slate'. In the example above, even though it has just compiled `greet.c` which contains the definition of the `hello()` function, this knowledge is 'forgotten' once it starts compiling `main.c`. Hence the warning. The *linker* is fed with the resulting object code from the previous compiles, and somewhere in there is the definition of `hello()`, so *it* can do its job and produce a runnable program.

As we mentioned before though, it is not a good thing to ignore this warning (or any warning, for that matter). We could add the prototype (see §3.5) of the `hello` function to `main.c`, making it look like so:

```
void hello(void);     /* defined in greet.c */

int main(void)
{
```

```
    hello();
    return 0;
}
```

For more complex projects, it is nicer to put this prototype in a header file. More on that in the next section.

10.2 Header Files Revisited

A good reason for putting prototypes in their own header files is that we can then keep related things together: a file greet.c contains the *implementation* of the function, and the file greet.h contains the *interface* to this functionality. We have been using a similar technique in most of the examples so far, by including system-provided header files such as stdio.h. In this section, we will look at creating such header files ourselves.

10.2.1 Your Own Header File

In effect, header files are rather simple. The only thing greet.h needs to contain is the prototype for the hello() function:

```
void hello(void);
```

The main.c program then looks like this:

```
#include "greet.h"

int main(void)
{
    hello();
    return 0;
}
```

Note that we don't use #include <greet.h> here, but instead enclose the filename in double quotes. This distinguishes *user* include files from *system* include files (*i.e.*, which are part of C or its standard libraries, or of the operating system).
It is wise to include the header file in greet.c as well, so that greet.c becomes

```
#include <stdio.h>
#include "greet.h"

void hello(void)
{
    printf("Hello, world!\n");
}
```

The reason for this is that otherwise, if you make a mistake in the implementation of a function, for example by defining it with the wrong number of parameters or with the wrong types, the compiler won't catch this. Since it hasn't seen any prototype when it encounters the function definition, it can't check whether the definition matches the prototype. Then, when compiling the code which *calls* the function (in a source file which does include the header file), it will trust the prototype and *assume* there will be a function somewhere which corresponds to it. Lastly, the linker will look for the function *by name*, ignoring its number and type of arguments.

Of course, this is all a little far-fetched for the simple `hello()` function, so let's add another. This one should print out a personalized greeting. Add the following function to `greet.c`:

```
void greet(const char *name)
{
    printf("Hello, %s!\n", name);
}
```

and then add its prototype to `greet.h`, making it look like

```
void hello(void);
void greet(const char *name);
```

We can now use this new function in our `main()`:

```
#include <stdio.h>
#include "greet.h"

int main(void)
{
    char name[80];     /* should be plenty */
    printf("Please enter your name\n");
    scanf("%79s", name);
    greet(name);
    return 0;
}
```

10.2.2 Error Prevention

Now, let's try to insert a few mistakes, and see how the header file helps prevent them. If we would simply call

```
greet();
```

we would get an error to the effect of `too few arguments to function 'greet'`. If we hadn't included the `greet.h` header (or didn't add the corresponding prototype to it), we would have gotten away with this. There might have been a warning about the implicit declaration of function `greet`, but it would compile, and the resulting code would be very wrong. You could even call it with a parameter of the wrong type, say '`greet(42)`' which is likely to crash! (The resulting program treats the value 42 as if it were the address of a string, and starts looking for one at that address. The value 42 is quite unlikely to be a valid pointer, so the operating system terminates the program for nosing around in memory which is outside its valid range.)

On the other hand, if we don't include the header file in the *implementation* file `greet.c`, we could make the mistake of defining the function with a different type of argument:

```
void greet(int n)
{
    int i;
    for (i = 0; i < n; i++)
        printf("Hello, world\n");
}
```

The caller tries to call the function with a `const char *`, but the implementation treats it as an `int`. If you're lucky, the value as it's found on the stack is interpreted as a small or negative number, but you might as will end up with a million greetings or more.

This mistake would not have been possible if `greet.c` included the proper header file containing the prototype, because it would have yielded an error:

```
greet.c:10: conflicting types for 'greet'
greet.h:2: previous declaration of 'greet'
```

In this case, the compiler is even being so good as to point out where it first saw the declaration of the `greet` function, so you can easily find what it *should* have been.

The examples here may look rather contrived. For small, one-person projects you are not very likely to run into trouble. However, errors like these are very common in large-scale projects, especially with multiple people working on the same project. In such cases, people are typically responsible for a certain subset of the source files. For example, you may be implementing the mathematical 'core' of a complex simulation, while someone else is working on its user interface (the part of the program which handles interaction with the user). In such cases, it is standard practice that everyone concentrates on getting the header files done first. These form a kind of 'contract' between the code calling the function and the code implementing it. It can be quite difficult to get this contract 100% right the first time, because new insights during the development of the program may require changes to the prototypes (for example, it may turn out the implementer needs an extra parameter here and there). As long as the header files are kept in sync, the compiler will pin-point the problematic areas for you.

10.2.3 Multiple Inclusions

Note that you are free to *declare* functions as often as you like (*i.e.*, it is not an error to include the same header file twice if it contains only function prototypes), as long is it is *defined* only once. If there do happen to be two definitions in the same source file, you'll get an error such as

```
greet.c:15: redefinition of 'greet'
greet.c:10: 'greet' previously defined here
```

If a function is redefined in a different source file (say, once in `greet.c` and once in `main.c`), this will only be found out at link time, giving an error like

```
greet.o: In function 'greet':
greet.o(.text+0x0): multiple definition of 'greet'
main.o(.text+0x2c): first defined here
collect2: ld returned 1 exit status
```

There are other things besides function definitions which must be seen only once by the compiler. If your header file happens to contain a `typedef`, for example, including it twice is an error (even though the two `typedef`s will obviously be identical, the compiler will say something like `conflicting types for 'yourtype'`). Now, including a header twice in the same source file may seem silly, but it can happen *indirectly*, if one header file includes another.

Take one of the database examples from chapter 6. A very reasonable split-up would be that one source file contains the functions for entering, searching, and removing entries (say, `database.c`), one source file contains functions for loading and saving an entire database from/to persistent storage (say, `storage.c`), and one contains the user interface code (say, `ui.c`, perhaps using a menu to present options to the user). Each of these functional parts of the program would have its own header file (`database.h`, `storage.h`, and `ui.h`, respectively). This presents us with a problem: Where do we put the definition of `person_t`? Each of the header files needs it (otherwise you'll get `parse errors` as soon as the compiler sees the declaration of a function which takes such an unknown `person_t` type as one of its parameters). One solution is to simply pick one of them. This means that the other header files need to `#include` this specific header file.

Another, more common solution is to put this `typedef` in a separate header file (say, `person.h`), and include this one in each of the other header files. However, as soon as one implementation file includes two or more header files (as typically happens in the implementation file containing the `main()` function, because there all the functionality 'comes together'), it will cause the '`conflicting types`' error, because the main source ends up indirectly including `person.h` three times!

There are two main 'schools of thought' regarding this issue.

The first one states that there needn't be a problem if you simply *disallow* multiple inclusions of the same header file. You are not supposed to include header files from within other header files. The 'end user' of your header file (someone who wants to use the functionality you provide and includes your header file in his source file) simply has to include all the other necessary header files himself. In other words: every *source* file using one of your functions starts with `#include "person.h"` before including the other header file.

The second one states that this places an unreasonable burden on the end user, because he has to remember to include this other file (you can imagine that in complex projects, there are several of these 'must include first' files). Failure to do so results in parse errors, and it is then up to the end user to figure out what file he should have included – perhaps by looking through all the header files to find out where a certain type is defined. This school of thought says that if a header file requires a certain type to be defined, it must include the relevant header file itself.

Clearly, the second school is more 'user friendly'. However, it still leaves us with the problem of multiple defined types. Somehow, we would have to figure out a way to prevent multiple definitions even if the corresponding header files are included more than once.

This can be achieved by using *header guards*, an interesting trick using the preprocessor.

10.2.4 Intermezzo: Conditional Compilation

We have been using only a few features of the C preprocessor so far, namely its function to `#include` other files into the one currently being processed, and its ability to `#define` macros. However, the preprocessor has a few more tricks up its sleeve. A very important one is to check whether a macro is defined or not, and compile different pieces of code depending on the outcome. For this, the preprocessor has the `#ifdef`, `#else`, and `#endif` directives:

```
#ifdef MACRO
    code which will be compiled if MACRO is defined
#else
    code which will be compiled if MACRO is undefined
#endif
```

For example, take the root-finding programs of §4.8. We added functionality to time how long it took to perform the calculations; something which not all users of the code would presumably be interested in. We can isolate these parts of the code and only compile them when a certain macro (say, `MEASURE_TIME`) is defined:

```
int main(void)
{
    double a = 0;
    double b = 2;
    double epsilon = 1e-7;

#ifdef MEASURE_TIME
    clock_t end;
    clock_t start = clock();
#endif

    /* root-finding algorithm here */

#ifdef MEASURE_TIME
    end = clock();
    printf("Found %f in %g seconds\n", (a + b)/2,
```

```
        (double)(end - start)/CLOCKS_PER_SEC);
#else
    printf("Found %f\n", (a + b)/2);
#endif

}
```

You can then compile a special version for performance measurements by simply adding a `#define MEASURE_TIME` at the top of your program. Most compilers even allow defining macros at the command line, so you don't have to modify your program at all, by doing something like

```
cc -Wall -DMEASURE_TIME -o root root.c
```

or, with Visual C++:

```
cl /W4 /DMEASURE_TIME -Feroot.exe root.c
```

10.2.5 Header Guards

We can use the ability of the preprocessor to tell whether a certain macro is defined or not to make sure the contents of a header file are only seen by the compiler once, regardless of how many times it is (indirectly) included. For this, we employ the `#ifndef` ('if not defined') preprocessor directive, which is like `#ifdef` but with its meaning inverted. Start each header file with

```
#ifndef FILENAME_H
#define FILENAME_H
```

and end it with

```
#endif
```

So for example, the first time the file `person.h` is included, `PERSON_H` will not be defined. When this is the case, it is immediately defined, and the normal contents of the header file follow. If the same file is included again later on, `PERSON_H` *will* be defined (by the previous encounter with the file), so the whole part between the `#ifndef` and the `#endif` (in effect, the entire file) is skipped.

Adding header guards to all your header files ensures the contents of the header files are only ever seen once during each compilation, meaning you don't have to worry about putting `typedef`s in them. It is heartily recommended you do this.

Since this trick is being employed on a wide scale, some compilers are even made aware of it. They 'remember' when a file contains header guards, and when they see a subsequent `#include` of such a file, they don't even bother reading it from disk a second time. In large projects, this can speed up compilation time significantly.

10.3 Translation Units

In §4.7, we introduced the concept of *scope*. Simply said, a scope is everything between { and }. Every function body is a scope, and you can nest scopes multiple levels deep. Anything declared in a scope is only visible within this scope.

There is also the *global scope*, which can be thought of as the scope of the entire program. Anything declared in this scope is visible throughout the code (any other scope can be seen as a nested scope of this global scope). In the example of the database programs of chapter 6, the code on page 133 puts the database itself and the `current_index` and `current_size` variables in the global scope.

However, now is the time to be a little more detailed as to what this global scope is. In effect, this scope is not for the entire *program*, but for the current *translation unit*. This is a fancy name for a source file (at the moment that all `#include`s have been processed, so as it is seen by the compiler).

This may seem a trivial distinction, but when you split up a program over several source files, something declared in the global scope of one source file is *not* automatically visible in the other source files! This was already apparent when it concerned functions and typedefs. These needed to be 'made visible' by bringing their declarations into the current translation unit (usually by means of an `#include` of the proper header file). However, with variables in the global scope this is different.

Consider the following example:

```
#include <stdio.h>

int global;

void f(void)
{
    global = 2;
}

int main(void)
{
    global = 1;
    printf("global is %d\n", global);
    f();
    printf("global is now %d\n", global);
    return 0;
}
```

If we split up this program, then the declaration of `global` ends up in either one of the files. That means compiling the other one will fail with an error to the effect of `'global' undeclared`.

We can refer to variables in one translation unit from within another by using the `extern` keyword. Suppose we chose to put the `int global;` in the source file containing `main()`, and put the `f()` function in a different source file, then we can still access `global` within `f()`:

```
void f(void)
{
    extern int global;
    global = 2;
}
```

Another option would be to bring this `global` variable into the global scope of the translation unit containing `f()`, by putting the `extern int global;` somewhere above the `f()` definition inside the same source file. There must be one translation unit where the `extern` is omitted. You can view `extern` as meaning 'Somewhere else, there exists a declaration of...'. If the compiler only ever sees `externs`, it will never generate the actual variable, and the linker will end up asking 'Okay, so where *is* it?'

As opposed to 'publishing' your functions and globals to all other parts of the program, you can also *limit* their visibility to the current translation unit. There can be several reasons for doing this.

One is a practical one: In a large project, especially when multiple people are working on different parts of the same program, you may run into 'name clashes' when you decide to write a small 'helper function', say `calculate()`, and someone else (in a source file far, far away) had that same idea. The linker will complain saying there are multiple definitions of `calculate`. This is called a *name clash*, and it is a nuisance, because you only ever intended to use this function yourself (perhaps as a step in an algorithm), and it is not part of your 'published interface'. However, even though you haven't mentioned this function in your header file, you still claim its name in the *global namespace*.

The other can be more 'philosophical'. You publish your interface in the header file, say for manipulating a database. Other users of your code, once they find out that there is a `database` array in the global scope, may 'fiddle around' with this, and manipulate the array directly instead of going through your functions. This may mean that certain assumptions you make in your functions (for instance, that the array is always sorted by name) may no longer be valid, and subsequent use of your functions may fail. In this case, you would like to prevent other parts of the program from accessing your 'internals' directly.

For these cases, you can use the `static` keyword. If you write

```
static person_t *database;
```

then the `database` pointer is only visible inside this translation unit. Similarly, if you do

```
static double calculate(double x)
{
    return x*x;
}
```

then you can only call this `calculate()` function from within the same translation unit. If someone else in a different source file happened to give a handy `x*x*x` function the same name, his `static` version of `calculate()` would only be visible to *him*.

Note that the `static` keyword also has a second meaning, which was explained in §8.3.2.

10.4 Building Revisited

10.4.1 Separate Compilation

Compilation time of complex projects can be dramatically reduced by making sure you only compile what *needs* to be compiled. In our example so far, we simply put all the source files comprising our program on a single command line, compiling and linking them all in one go. However, if only a single source file has changed, you only need to recompile this one file. You can then relink the resulting new object code, along with the unchanged object code for the other file(s), into a new executable.

For this, it is necessary to split up the compile and link steps. Taking the `greet` example from the start of this chapter:

```
cc -c -Wall main.c
cc -c -Wall greet.c
cc -o hello greet.o main.o
```

The first two lines perform the actual compilations. The `-c` flag means 'compile only, don't link'. The two resulting object files, `main.o` and `greet.o`, respectively, are then linked together with the third command.

On a Windows system with Visual C++, the examples above would look like

```
cl /c /W4 main.c
cl /c /W4 greet.c
cl /Fehello.exe greet.obj main.obj
```

Note that in both examples, we are using the command usually used for the compiler itself as a 'front end' for the linker. For example, in the Visual C++ case, we could also have used

```
link /out:hello.exe greet.obj main.obj
```

instead of the last line.

If you now modify only `main.c`, you can rebuild your program by only compiling `main.c`. The object file for `greet.c` is still around, so you re-use it for your updated program:

```
cc -c -Wall main.c
cc -o hello greet.o main.o
```

For small programs, this may seem to be more trouble than it's worth. But for really big projects it can mean the difference between a few hours to get a new executable with your changes incorporated, and a few seconds.

10.4.2 Dependencies

In the previous subsection, it was shown that you need not recompile *every* source file of your program when *one* has changed. This is an important reason for splitting up the source in the first place (to speed up compilation time). The exact wording was 'compile what *needs* to be compiled'. However, what exactly *needs* to be compiled?

If only the *implementation* of a function is modified (for example, a faulty calculation is replaced with the correct one, or a more efficient algorithm is chosen), the code which *calls* the function need not be recompiled.

When the *interface* changes, then obviously the caller needs to be modified and recompiled as well. When an extra parameter is added to the function prototype, you can imagine that this needs to be added in the code which calls the function as well.

But even when the difference is more subtle, for instance because one of the parameters of a function is a `struct` and the definition of this `struct` is changed, the caller needs to be recompiled – even if the actual source code is unchanged because the definition of the `struct` happens to be in some header file.

This turns out to be quite a nasty issue, because you need to take into account *all* header files, even those included indirectly. In the database example of the previous section, the definition of the actual `person_t` data type was moved to its own header file. Suppose this `struct` is changed to allow for a smarter implementation of some of the functionality in the `database.c` file, then even source files which include `database.h` need to be recompiled, and also the source files which include `storage.h` and `ui.h`. This is because all of these include `person.h` indirectly.

The proper terminology here is that all of these source file have a *dependency* on `person.h`. The problem is that tracking these dependencies can be quite a chore, and since dependencies can *change* when parts of the program are modified (because someone includes a header file which previously wasn't needed), they need to be evaluated every time you recompile.

You can imagine that in complex projects, doing this 'manually' every time you recompile costs more time than you save by doing partial recompilation in the first place, and doing it wrong can result in an incorrect executable.

10.4.3 Automated Build Tools

Luckily, there exist programs which can assist you here. We will look at two examples: `make` (which is often used on UNIX systems) and Microsoft Visual C++. If you are using the latter, you can skip to the last few paragraphs of this subsection.

The program `make` uses so-called *makefiles*. A makefile is a text file with 'rules', basically listing *targets*, *prerequisites*, and *commands*, and the `make` program uses this to decide what needs to be done when it is asked to rebuild a certain target. For the 'hello world' example we've been using in this chapter, the makefile would look like this:

```
hello: main.o greet.o
	cc -o hello main.o greet.o
```

In this example, `hello` is the target. It is followed by a colon, and then a listing of the prerequisites (*i.e.*, what is needed to build the target). Underneath, you list the command (prepended by a tab – not spaces!) saying *how* to build the target.

If you now type `make` at the command line, it will look for a file called `Makefile` (note the capital M) in the current directory, and if no specific targets are given on the command line, it will build the first one listed in the makefile. The `make` program has some rules built in, so it knows how to compile `.c` files into `.o` files. It will do this if the `.o` files are missing or it detects that their corresponding `.c` files have changed (it does that by looking at the *timestamp* of the files, noting whether the `.o` file is 'older' than the corresponding `.c` file).

Then, if any of the `.o` files listed as a prerequisite for the `hello` target have changed, it will perform the command listed in the makefile (*i.e.*, link them into an executable).

If you want to override the default action used for compiling `.c` files into `.o` files, you can specify a general rule:

```
.c.o:
	cc -Wall -c $<
```

This rule roughly translates as 'when you need to make a `.o` file out of a certain `.c` file, use the following action.' In the action, the `$<` token means 'whatever the full name of the `.c` file was'. In this case, we use this custom rule to switch on the maximum warning level for each compilation.

The example above lists only the dependencies of the final program, but each of the `.o` files have their own dependencies too. You can extend the makefile as follows:

```
hello: main.o greet.o
	cc -o hello main.o greet.o

.c.o:
	cc -Wall -c $<

main.o: main.c greet.h
greet.o: greet.c greet.h
```

Note that you don't *need* to specify a command: In this case, we only listed a few more dependencies, and the default action (which we happened to specify a few lines above) will be taken. You can verify that this makefile works by changing one of the source files and typing `make` (it will print out what it's doing). Tip: on UNIX systems, there is the command '`touch` *filename*' with which you can 'pretend' that you edited a file. It will simply set the timestamp to the current time, so `make` will notice that a source file is newer than its corresponding `.o` file, and recompile.

You should find that touching either `main.c` or `greet.c` will make `make` recompile only this source file (and subsequently relink the executable), whereas when you touch `greet.h` it will recompile both.

Most compilers offer a mode in which they generate the list of dependencies for you, so you don't have to track them manually. For GCC, this is achieved using the `-MM` command line option. If you type

```
cc -MM -c main.c
```

it will print out

```
main.o : main.c greet.h
```

There are ways to automatically generate the dependencies and include them in the makefile, but this is outside the scope of this book. There are whole books written about makefiles, and the makefiles for complex projects can be hundreds of lines long.

People using an Integrated Development Environment (IDE), such as Microsoft Visual C++ or Apple's Xcode, arguably have an easier time. Usually, such IDEs work with so-called *projects*. You simply add a new source file to the project, and all dependencies will be figured out automatically.

Rebuilding the program is usually just a single menu selection or hot-key; the IDE will save all the source files that were changed, recompile what needs to be recompiled, and relink the executable. Often, there is an option to start the program automatically once it's rebuilt.

The exact menu options and hot-keys vary, but there are usually options to add existing source files to the project, or open a new file which is automatically added. The name of the resulting executable can usually be specified in a 'settings' screen (in MSVC, it's one of the *properties* of a project) along with dozens of compiler/linker options (for optimization, target processor type, directories in which to look for header files, etc.)

The compiler and linker output is usually shown in a separate window in the IDE, and clicking on an error or warning opens the corresponding source file and conveniently puts the cursor right at the offending spot in the code.

The editor usually offers *syntax coloring* in which recognized keywords get a different color, some form of *auto-completion* in which typing '`avery`' lets the editor guess that you wanted to type '`averylongvariablename`' (if you happened to use this name before), and an integrated help system in which you can highlight a keyword and get documentation at a keypress. For example, when your cursor is on a `printf` function call in your code, pressing `F1` (the usual choice for the 'help key') might give you a page about `printf`, describing all the format specifiers, what the return value means, and which header file to include to use this function in your program.

Many sophisticated IDEs also contain a *debugger*, which is a tool which lets you step through the source code of a program *as it is running*, so you can track what the program is doing, skip certain lines of code, inspect the current value of variables, or even modify them. In some, you can even suspend program execution, change some code, rebuild, and continue running the modified program!

Using IDEs can speed up your programming work considerably, and many of their features can be rather 'addictive', in that once you're used to them, going back to a development environment without them can be very frustrating. However, it is important to realize that in the end, you still generate *plain text files* with your source code in them, and behind the scenes there is still a compiler translating them into object code, and a linker building an executable out of them. The source code files can be moved to a separate computer with a different compiler installed on it, and (provided you didn't use any platform specific functionality in your program) be compiled and run there as well.

10.4.4 Include Paths

When header files are located in the same directory as the source files including them, they are automatically found by the compiler. For very large projects, or projects using libraries (see the next section), header files may be located in different directories. You can specify to the compilers what directories to look in. On UNIX systems, this is usually done with a `-I` command line switch, as in:

```
cc -o myprogram -I/path/to/headers mycode.c
```

With Visual Studio, the switch is `/I`*path*. You can also add directories in the settings of your project file, if you are using the Visual Studio IDE.

10.5 Libraries

When you have been programming for a while, you have probably acquired a small collection of 'handy functions', which were useful to have around in many of your assignments or projects. Things like `wait_for_enter()` perhaps. Before this chapter, you may have simply 'pasted' them into your current project. If this is indeed the case, then you have undoubtedly been thinking about these functions in light of the subject matter of this chapter. It would be a nice solution to keep all of these functions together in a single source file, say `utilities.c`, and simply compile this file along with all your other projects.

But why not go a step further? Since this 'library' of useful functions doesn't change much, you can keep a pre-compiled object file around, and simply *link* it into your future programs. When you do this, you are effectively providing a *library* (albeit for personal use only).

In fact, any time you have implemented some functionality which is not *specific* to your current program, you may want to group it together and keep a pre-compiled library of it around. You can link multiple object files together into such a library. To the linker, the only real difference is that it doesn't link in the 'glue' to make it an executable file. You can then later link this library into your other programs.

An example would be the database code from chapter 6. You can imagine that having database-like functionality could be handy in several different programs, and you could link all the code mentioned there (except for the `main` function) into a library.

Another example is the `next_permutation()` function on page 169. Most of the code in the 3D puzzle example of chapter 8 was specific to the example, but this particular function was made *generic* (it takes the size of the array as a parameter even though it will only ever be called with n = 6 in the 3D puzzle program). In many cases, it is not (much) more work to write a function in a generic way. That could make it useful in completely unrelated programs, and thus a good candidate for inclusion in a library.

On the other hand, the function `print_configuration()` on page 175 was very specific to this particular 3D puzzle solver. Making it generic (taking a variable number of puzzle pieces and supporting different layouts, for example) would take considerable effort, with questionable 'return on investment'.

There is a lot to be said about libraries, and there exist hundreds of libraries for all kinds of functionality, from reading and writing files of a certain type and implementing data compression to calculating Fourier transforms or displaying graphical elements on screen. Chapter 12 focuses entirely on libraries.

10.6 Callbacks and Function Pointers

A special case of 'interface based programming' is when you use functionality provided elsewhere (in the same program, or even in a library), but the implementation of this functionality in turn needs functionality provided by you. For example, suppose you use a library which can sort a given array of data in ascending order. If this array contains simple numbers, determining when two elements are 'in ascending order' is easy for the library. But what if the array contains some `structs` which you defined yourself, and which the library cannot possibly know about? You would like to use the *algorithm* implemented in the library, but provide your own *comparison function*. This is possible through a mechanism called *callbacks*. In §5.5, it was briefly mentioned that C has the concept of 'pointer-to-a-function', and this is a situation where they come in handy.

The standard C library happens to contains a function to sort an array in ascending order. Its prototype (in `stdlib.h`) is:

```
void qsort(const void *base, size_t nmemb, size_t size,
           int (*compar)(const void *, const void *));
```

This sorts an array of `nmemb` elements pointed to by `base`, each of which is `size` bytes. The last (funny looking) parameter tells us that the `compar` argument is a pointer to a function taking two `const void` pointers and returning an `int`. For each comparison the `qsort` (for QuickSort) algorithm needs to make, it will call that function with pointers to the two elements in the comparison. Your function is expected to return zero if the elements are identical, minus one if the former is 'less than' the latter, and plus one if the former is 'greater than' the latter (for your particular definition of 'less than' and 'greater than'). The arguments to the comparison function have to be `void` pointers, because the library cannot know about your particular data type. You will have to cast to the 'real' data type in the implementation of your function.

As an example, suppose you want to sort an array of the `person_t` database records from §6.3.4. Your comparison function then may look something like this:

```
int compare_names(const void *a, const void *b)
{
    const person_t *pa = (const person_t *)a;
    const person_t *pb = (const person_t *)b;
    return strcmp(pa->name, pb->name);
}
```

You can then call the `qsort` function like so:

```
qsort(array, size, sizeof(person_t), compare_names);
```

The nice thing here is that you can change the criterion which you sort the array by, simply by passing a different comparison function. That way, you can easily sort the list of `person_t` entries by address instead of by name.

Note that there is no special syntax involved in passing a function as an argument. C allows (but does not require) prepending an 'address-of' operator:

```
qsort(array, size, sizeof(person_t), &compare_names);
```

which some people like because it makes it clear that you are passing a *pointer* to a function.

Another situation where callbacks are often used is in *event driven programming*, which is common in interactive programs. Many libraries (some provided by the Operating System) used for designing graphical user interfaces allow you to register callbacks (also called 'event handlers') which are to be called when a user clicks a certain button on the screen, or selects a certain menu item, for instance.

The syntax for *declaring* a function pointer is something which even seasoned C programmers often need to look up. It is:

return_type `(*name)(` *type_1* `param_1,` *type_2* `param_2, ...)`

It is usually helpful to use a `typedef`:

```
typedef int (*compare_func)(const void *a, const void *b);
```

Inside the body of the `qsort()` implementation, the `compar()` function can be called by dereferencing the pointer to it:

```
    if (!(*compar)(&itemA, &itemB))
    {
        // itemA and itemB are identical
    }
```

Since `compar` is a pointer-to-a-function, `*compar` is the function itself. You can also put the 'real' type in the comparison function, like so:

```
int compare_names(const person_t *a, const person_t *b)
{
    return strcmp(a->name, b->name);
}
```

and use the cast in the call to `qsort` instead:

```
qsort(array, size, sizeof(person_t), (compare_func)compare_names);
```

This is better, because it makes more clear what the `compare_names()` function expects without having to look into its implementation.

10.7 Synopsis

Programs in C need not consist of a single source file. The source code for a single program can be split up in several source files. This is done to make large projects easier to maintain and to reduce the time needed to rebuild the project in the face of changes.

To call a function in one source file which is defined in another, the function *declaration* (or *prototype*) needs to be 'seen' by the compiler. Prototypes and type declarations are usually placed in *header files*. Another view of this is that the *interface* is in the header file, and the *implementation* in the source file.

Separating out common type and function declarations into their own header files can lead to problems when the same header file is included multiple times (indirectly, via other header files). This can be solved by using the *header guards* technique. This uses the C preprocessor feature of *conditional compilation*, in which a part of a source file is either compiled or skipped based on the (lack of) definition of a preprocessor macro.

A source file with all its header files included, *i.e.*, as it is seen by the compiler, is called a *translation unit*. The global scope of one translation unit is not visible by default in other translation units (although the linker will complain if multiple translation units contain definitions for the same symbol), but specific entities can be 'imported' in another translation unit by using the `extern` keyword. By using the `static` keyword, you can restrict the visibility of functions and variables to the current translation unit. In that case, other translation units can't 'see' this symbol, not even using `extern`. This is useful to prevent *name clashes*.

The `static` keyword can also be used to save the value of a variable local to a function across calls to this function.

Source files can have *dependencies* on other files, usually header files. This means that when a certain header file is changed, all source files having a dependency on this file (even indirectly) need to be recompiled when rebuilding the program. There exist automated build tools to handle the chores of figuring out which files need to be recompiled.

C allows you to pass a pointer-to-a-function as an argument in a function call. This enables the use of *callbacks*.

10.8 Other Languages

The ability to split up the source code of a program in amenable 'chunks' is a tell-tale sign of whether a programming language was designed for use in large, complex projects. Most modern programming languages allow it; some use the term *units* or *modules*.

In some 'older' languages, for example BASIC, a program was always just a single (long) listing of statements. Since the B in BASIC stands for 'beginners', this is not surprising.

For interpreted languages or scripting languages, the 'separate compilation' argument is not applicable, and they are rarely intended for large projects anyway. As was mentioned before, in many scripting languages syntax errors are only found when the program reaches the offending statement (because there is no compiler which has to 'approve' the entire program first). This alone is a strong argument against using them for complex projects, because it quickly becomes undoable to guarantee that your testing of the program has touched *every* line of code in it.

In object-oriented languages (see §6.5 for a brief overview), an 'object' can be viewed as an intermediate level between a function and a translation unit.

The problems mentioned in §10.2.2 are rather specific to C (and C++). Other languages which support modular development usually provide in a more 'robust' way of making sure the caller and the callee are matched.

The concept of *interface based programming* has proven very fruitful in the development of large, complex programs. It is a natural match for object oriented programming languages, because you can have several objects implementing a certain interface, and as a 'client' of these objects, all you care about is this interface. However, you can use similar techniques in C as well. For example, on the Microsoft Windows platforms, a technique called COM (Component Object Model) is very popular. This is an interface and object oriented technique where the interfaces are specified in a special (C-like) language called IDL (Interface Description Language), and objects implementing these interfaces can be written in several languages (also, albeit a little laboriously, in C).

Libraries are often available in other languages as well. For scientists, it is useful to note that there is a huge number of libraries available for FORTRAN implementing all kinds of numerical algorithms. Oftenm, there is a way to call these functions from C.

The concept of *callbacks* is available in many other languages in a more refined form, such as *signals and slots* in C++, and *events and delegates* in C#.

10.9 Questions and Exercises

10.1 Experiment with Makefiles or projects, depending on the system you're working on, by making modifications in part of a program and noting which files get rebuilt. Especially try to get 'nested dependencies' where header files depend on other header files.

10.2 In a 'do not try this at home' experiment, investigate what happens when you do *not* use automated dependency tracking, and end up with a program which calls a function with a type of argument different from which it expects. Especially interesting are mix-ups between pointer and 'normal' types. Can you get your program to crash? Even if compilation gave no errors?

10.3 If you are using an IDE, try to run one of your programs in the *debugger*, after putting a *breakpoint* on the very first line of your `main` function (see the documentation of your IDE for help on how to do this – there is often a menu item in a 'debug' menu to put a breakpoint on the line which the cursor is currently on). Then, execute your program step by step. If it contains function calls, many debuggers offer options to either 'step into' a function, or execute the function call and continue on the next line. Experiment with this.

10.4 ⋆ The shift from 'implementation-oriented programming' to 'interface-oriented programming' is often difficult to make, and many programmers switch back and forth between interfaces and implementations during the development of a program without realizing it. To become more aware of this, it helps to find a partner and tackle a relatively large assignment *together*, while trying to minimize the amount of communication. In a first session, you should try to avoid talking about *how* to solve particular sub-problems, but instead focus on *what* needs to be solved. Come up with an *interface* in the form of header files; one person then goes off to implement the functionality, and the other person implements the 'client code' which makes use of this functionality.

A suggestion for a 'relatively large assignment' is to write a program which takes a chemical reaction equation and fills in the correct number of molecules on both sides. With reaction formulas, you often begin with noting which compounds get converted into which other compounds, and figure out how to make the numbers match in a second step. For example, when burning ethanol, there are molecules of ethanol reacting with molecules of oxygen, yielding carbon dioxide and water. To make the numbers of atoms left and right of the arrow the same, we need a certain number of molecules:

$$C_2H_5OH + \mathbf{3}O_2 \rightarrow \mathbf{2}CO_2 + \mathbf{3}H_2O$$

Write a program to fill in these numbers (bold-faced in the equation above), given the reagents and the reaction products. This looks deceptively simple, but there are quite a few subproblems to solve here. Figuring out a 'natural syntax' so the user can enter which compounds react to yield which other compounds, and a way to keep track of the number of atoms, is perhaps more difficult than the actual algorithm to calculate the minimal number of molecules of each type to make the numbers match.

Chapter 11
Graphics

> By the time the child can draw more than a scribble, by the age of four or five years, an already well-formed body of conceptual knowledge formulated in language dominates his memory and controls his graphic work. Drawings are graphic accounts of essentially verbal processes. As an essentially verbal education gains control, the child abandons his graphic efforts and relies almost entirely on words. Language has first spoilt drawing and then swallowed it up completely.
>
> *Karl Buhler*

11.1 Graphical User Interfaces

If you have ever used a computer after the mid eighties of the previous century (which is not unlikely if you are reading this book) then you must by now have wondered something about the appearance of the programs treated so far. They all produced output (some after taking some input) but all in a rather 'textual' way.

If you ask a random person on the street to describe what a computer program 'looks like' in the most generic terms, he or she will probably come up with a description featuring 'screens' or 'windows', 'buttons you can click' etc. The way computers are operated has changed dramatically since the advent of the so-called 'graphical user interface', or GUI for short.

GUI programming, however, is not standardized for C (nor for many other languages). It is often a feature of a particular operating system, or of a library (see the next chapter) at best. Of course, C is a perfectly capable language to write graphical applications in. For example, the performance-critical parts of many computer games (some of the most demanding graphic applications on todays computers!) are written in C.

But the C standard says nothing about graphics or GUI programming. In fact, the only occurrence of the word 'graphic' in the C99 standard is when describing the character set available to C programmers; *graphic characters* are non-letter, non-digit characters such as `!`, `<`, `{`, and `%`.

In contrast, some programming languages have graphics primitives built in (for plotting points on the screen, drawing lines, etc.), or even an extensive GUI-building toolkit. An example of the latter is Visual BASIC, in which the programmer can visually lay out a dialog window with buttons and other controls. It is then a matter of filling in the 'handler functions' which get called when the user clicks a particular button (or selects a particular menu item, etc.) and the GUI is finished. Usually, such a programming language is tied to a certain platform, although there are cross-platform GUI building tools and libraries available.

For many scientific programs, a 'command line interface' is perfectly adequate. After all, there is not much use in writing a GUI which lets you enter some initial value in a text input field and which displays the resulting numerical output in an 'alert box' after you click a 'Calculate!' button.

On the other hand, even the 'pseudo-graphical' output from the programs of chapter 9 is a far cry from the typical computer program running on your own computer. For some areas, having graphics is a real benefit, and we will examine some of the concepts in this chapter.

11.2 Visualization vs. Interpretation

Although the first thing which comes to mind when thinking about computers and graphics is the *production* of graphics, there is a whole area of computer science concerned with programs taking images as *input*. This is called *image processing*, and its aim is either to enhance images (to make them more 'appealing' to humans or so humans can interpret them more easily) or to extract data from them (so humans need not look at the images at all). Examples of 'enhancing' range from red-eye removal to contrast stretching or lens distortion correction. Examples of data extraction range from finding the size distribution in an image of particles to automatic recognition of faces.

We will not treat image processing here, but we will investigate some aspects of data visualization. Unfortunately, due to the fact that the available operating system or library functions (see the next chapter) for putting graphics on the screen vary wildly from platform to platform, it is not feasible to examine 'interactive' graphics here. In fact, it turns out that the best we can do to produce an image is to write the corresponding data to a file, and rely on the operating system to display the contents of this file on screen.

Using this 'clumsy' workaround, it is relatively easy to make rather impressive graphics with little effort, and in a portable way.

Of course, there are other approaches to visualizing your data. If all you need is a graph of some array of data your program produces, you can write this data to a file and use an external program to render a graphic representation of the data. An excellent data plotting program is gnuplot, which is a command-driven interactive function plotting program. It is available for a multitude of platforms (including Windows, Mac OS X, and Linux). It can be used to plot functions and data points in both two- and three-dimensional plots in many different formats. Gnuplot is copyrighted, but freely distributable; you don't have to pay for it. Gnuplot can be downloaded from `http://www.gnuplot.info`.

As an example, below is a figure generated with gnuplot, with the output of the swinging pendulum simulation with air friction of §9.2.3.

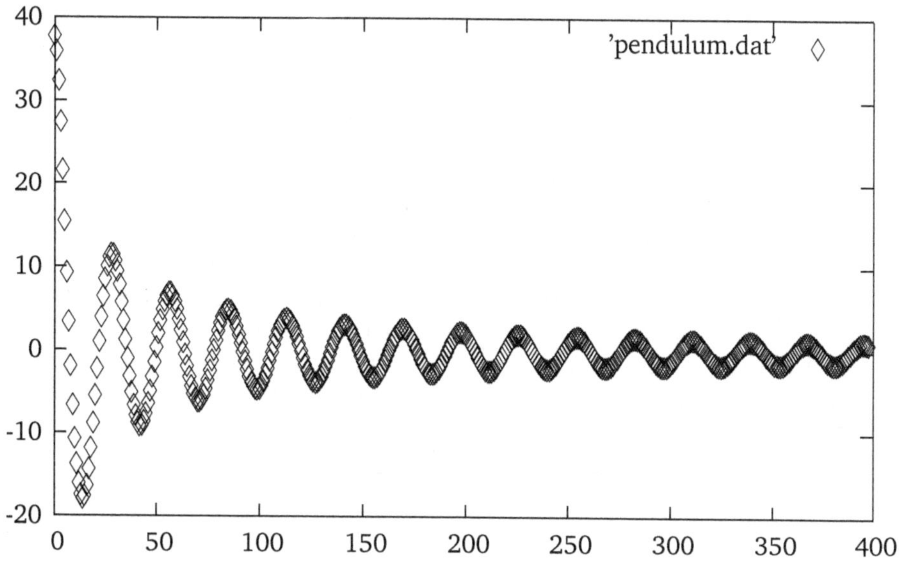

11.3 Vectors and Bitmaps

In computer graphics, usually the distinction is made between *vector graphics* and *bitmap graphics*. In the former, a graphic is described in terms of *primitives*, such as 'red line going from coordinate (10, 32) to coordinate (20, 30)' or 'blue rectangle of 20 by 40 units, with its left top located at coordinate (100, 50)'. Vector graphics are *resolution independent*: in principle, the coordinates can be expressed in floating point.

In bitmap graphics, an image is viewed as a (large) matrix of *pixels* ('picture elements') which each have a certain color. This is how, for example, a photo can be represented on a computer. When these pixels are small enough and/or the range of color values representable on the output device is large enough, the human eye usually does not discern the individual pixels.

In fact, most output devices for computers render in a pixel-based fashion, with the exception of *plotters* which basically consist of a pen mounted on an x,y stage. When rendering a vector-based graphic to a printer or computer screen, it is first converted to a bitmap.

There are several ways to encode color values in a bitmap. The pixels in the bitmap can contain color values directly, for example as a gray value encoded as a number (*e.g.*, 0 for black, 255 for white, and everything in between as gray values), or in a triplet of red, green, and blue values (using one byte for each, as is usually the case, this yields a spectrum of $8^3 = 16\,777\,216$ possible colors). Whereas the monitors use this RGB (red, green, blue) model, there are other color models as well, such as CMYK (cyan, magenta, yellow, and black ('key'), which are the different colors of ink used in four-color printing presses). Bitmaps can also use a 'level of indirection' where each pixel value is actually a 'color number' to be looked up in a *palette* which contains the real color value.

11.3.1 GUIs Revisited

Since a computer screen is in essence a matrix of pixels (usually with dimensions like 1280×1024 or 1440×1050 or so), the entire contents of the screen can be viewed as a bitmap too. Its contents are controlled by the operating system. When your program uses `printf()` to print a certain string to its window, the operating system iterates over each of the characters in this string, retrieves small bitmaps for each of the glyphs (or first renders those small bitmaps out of vector representations of that glyph) and copies those bitmaps to the 'screen bitmap' (called the *frame buffer*) at the correct location.

Programs with a graphical user interface can likewise request the operating system (via specific library calls) to draw a line or rectangle for example, or put a representation of a button on a specified location on the screen.

11.3.2 Image files

It was mentioned in §7.4.3 that there are several *file formats* (standards for storing information in files) pertaining to graphics. Some are vector oriented (such as PostScript), but we will focus on bitmap files here, particularly the BMP format, which is a standard bitmap file format on the Windows platform. This is certainly not the most elegant nor the most simple format, but support for it is available on a large range of platforms (including Mac OS X and UNIX systems).

Making a function to *write* BMP files is considerably easier than one to *read* them, because the latter would have to handle all possible varieties of BMP files (of which there are quite a few) whereas a BMP writer only needs to make sure that its output conforms to *one* valid type of BMP files. In our case, we will use the '8 bit palette' variety, where the file contains a general information header (with the dimensions of the image, the type of BMP, etc.), followed by an array of color values, and finally the image itself in the form of an array of pixels.

BMP files are *binary files* (as explained in §7.4.2), which means that we need to take special precautions to make sure that numeric values are written into the file in the correct 'byte order'. In this case, the BMP file format requires 'little endian' order (the native order on Intel x86 systems). To make sure our code also works on PowerPC systems, we will use special functions to write multi-byte values into the image file header:

```
void write16_LE(char *p, short a)
{
    p[0] = a%256;
    a /= 256;
    p[1] = a%256;
}

void write32_LE(char *p, long a)
{
    p[0] = a%256;
    a /= 256;
    p[1] = a%256;
    a /= 256;
    p[2] = a%256;
    a /= 256;
    p[3] = a%256;
}
```

These functions write 16-bit and 32-bit values, respectively, making sure that their LSB (least significant byte) is written in the first memory location.

A BMP header is 54 bytes in size. For a precise description of what all the bytes in the header mean, google for 'BMP file specification'. For now, we will simply continue with a function which writes a valid BMP file header for an image of given dimensions (w×h pixels), followed by the color table (or palette). Each entry in the color table is a four-byte value, containing the values for the blue, green, and red components (with one byte 'spare'). In the code, the meaning of some of the values is given in a comment. An important thing to note is that each line in a BMP *file* needs to be a multiple of 4 bytes. If the width of the image itself is not a multiple of 4 pixels, each line needs to be *padded*.

```
typedef char color_t[4];

void write_BMP8_header(FILE *fp, int w, int h, color_t palette[256])
{
    char header[54];
    int padw = ((width + 3)/4)*4;
    long filesize = sizeof(header) + 256*sizeof(color_t) + padw*h;

    /* BMP files start with these two 'magic' bytes */
    header[0] = 'B';
    header[1] = 'M';
    write32_LE(header + 2, filesize);
    write16_LE(header + 6, 0);
    write16_LE(header + 8, 0);
    write32_LE(header + 10, 54 + 256*sizeof(color_t));
    write32_LE(header + 14, 40); /* remaining size of header */
    write32_LE(header + 18, w);
    write32_LE(header + 22, h);
```

```
    write16_LE(header + 26, 1);  /* one image plane */
    write16_LE(header + 28, 8);  /* 8 bits per pixel */
    write32_LE(header + 30, 0);  /* no compression */
    write32_LE(header + 34, 0);  /* file size not specified */
    write32_LE(header + 38, 0);  /* x resolution (default) */
    write32_LE(header + 42, 0);  /* y resolution (default) */
    write32_LE(header + 46, 0);  /* #colors = 256 */
    write32_LE(header + 50, 0);  /* all colors important */

    fwrite(header, sizeof(header), 1, fp);
    fwrite(palette, sizeof(color_t), 256, fp);
}
```

As an example, the function below prepares a palette which contains all possible shades of gray which can be represented in a BMP file (going from entirely black to entirely white), by setting the red, green, and blue components for each palette entry to be equal:

```
void fill_palette(color_t palette[256])
{
    int i;

    for (i = 0; i < 256; i++)
    {
        palette[i][0] = i;
        palette[i][1] = i;
        palette[i][2] = i;
        palette[i][3] = 0;
    }
}
```

Next, we'll define our own 'image' data type, along with functions to initialize and clean up:

```
typedef struct
{
    int width;
    int height;
    char *pixels;
} image_t;

void init_image(image_t *img, int width, int height)
{
    img->width = width;
    img->height = height;
    img->pixels = (char*) malloc(width*height);
}
```

```
void deinit_image(image_t *img)
{
    free(img->pixels);
}
```

Finally, we'll look at a function to write out the entire image to a file, taking care of writing out extra padding if necessary. Since BMP files are 'upside down', *i.e.*, the first pixel in the file is the bottom left pixel instead of the top left pixel, we'll write out the lines in reverse order:

```
void write_BMP8(FILE *fp, image_t img, color_t palette[256])
{
    int y;
    int npad = 0;
    char pad[4];

    if (img.width%4)      /* we need padding */
        npad = 4 - img.width%4;

    write_BMP8_header(fp, img.width, img.height, palette);
    for (y = img.height - 1; y >= 0; y--)
    {
        fwrite(img.pixels, 1, img.width*img.height, fp);
        fwrite(pad, 1, npad, fp);
    }
}
```

Now, let's test these functions by writing a small BMP file containing a nice 'gradient' (color ramp) of all available colors in the palette. Running the program below should result in a file `test.bmp` which you can view using your operating system's standard image viewer.

```
int main(void)
{
    color_t palette[256];
    image_t img;
    int x, y;

    FILE *fp = fopen("test.bmp", "wb");
    if (!fp)
    {
        printf("Couldn't open test.bmp for writing\n");
        return -1;
    }

    init_image(&img, 200, 256);
    for (y = 0; y < 256; y++)
        for (x = 0; x < 200; x++)
            image.pixels[y*200 + x] = y;
```

```
        fill_palette(palette);
        write_BMP8(fp, img, palette);
        fclose(fp);

        deinit_image(&img);

        return 0;
    }
```

When doing image processing in C, it is custom to use a single allocation for the entire image. This is why the `image` array is *one*-dimensional, and not 2D as you might expect. Declaring it as

```
        char pixels[256][200];
```

(see §5.4) would make addressing a specific pixel easier as the pixel at (x,y) would simply be `pixels[y][x]`, but if the image dimensions are not known at compile time (which they often aren't), allocating the image this way would be very clumsy as we would have to use an array of *pointers* to lines of pixels, and each line would have to be allocated (and freed) separately in a `for`-loop.

Note that a pixel at (x,y) can be addressed in the 1D `image` array by calculating `y*width + x`. Of course, we can use a function lie this:

```
    void set_pixel(image_t *img, int x, int y, unsigned char value)
    {
        img->pixels[y*img->width + x] = value;
    }
```

See exercise 11.6 as to why this is usually not done if *all* pixels of a bitmap need to be set.

11.4 Two Useful Examples

Now that we have a way of displaying bitmaps on screen, let's look at two examples of interesting bitmap content.

11.4.1 Dithering

Let us define a function to draw (straight) lines into a bitmap, so we can draw simple shapes:

```
    void draw_line(image_t *img, int x0, int y0, int x1, int y1,
                   unsigned char value)
    {
        double len = hypot(x1 - x0, y1 - y0);
        double dx = (x1 - x0)/len;
        double dy = (y1 - y0)/len;
```

```
        double x = x0;
        double y = y0;
        int i;

        for (i = 0; i < len; i++)
        {
            int ix = (int)(x + 0.5);
            int iy = (int)(y + 0.5);
            img->pixels[iy*img->width + ix] = value;
            x += dx;
            y += dy;
        }
    }
```

This function draws a line into the image from (x0, y0) to (x1, y1), setting the pixels to value. It does this by first calculating how long this line is (len), determining how big the steps are which need to be taken in horizontal and vertical directions (dx and dy), and then adding these steps to the x and y coordinates for each position along the line.

Since x and y are incremented with fractions of a pixel for each step, they are doubles[1]. Of course, to calculate the corresponding offset into the pixel array, they have to be converted to integer. Adding 0.5 ensures proper rounding (since the (int) cast simply truncates). Replacing the two nested for-loops in the program on page 237 with the following code will draw a simple house-like shape:

```
draw_line(&image, 100, 50, 100, 250, 0);
draw_line(&image, 100, 250, 200, 350, 0);
draw_line(&image, 200, 350, 300, 250, 0);
draw_line(&image, 300, 250, 300, 50, 0);
```

If you experiment a bit with this draw_line() function to draw a few shapes yourself, you may notice that lines which are perfectly horizontal, perfectly vertical, or at around 45° look fine, but lines which are 'nearly but not quite' horizontal or vertical (for example, a line going from (100, 50) to (105, 250)), don't look 'smooth'. Since they are essentially composed of straight lines, the illusion of a diagonal line is disturbed and you notice a 'staircasing' effect. In computer graphics, this effect is known as *aliasing*, and its basic cause is that pixels are not infinitely small, nor can they be 'partially colored'.

The effect is shown below:

There are two common remedies for this. One is called *anti-aliasing* and uses gray values to simulate 'partially colored' pixels. Although we do have gray values available, we will look at a different approach which is also useful in other areas, namely *dithering*.

[1]There are also algorithms for line-drawing which are entirely integer-based, such as the Bresenham line algorithm.

To understand the concept of dithering, consider the following analogy. Suppose you go to the bakery for a loaf of bread. The loaf costs $1.60, but you only have paper money on you, so you hand over $2. Unfortunately, the baker is out of change. You agree that he owes you 40 cents, and you go home. Same story next time: both of you only have 'integer dollars'. You decide that since the baker owes you 40 cents, you get a loaf of bread for just one dollar, and you now owe the baker 20 cents (assume the baker knows you and trust that you will not keep the money and flee to a tropical island with it). Next time, you pay $2 again; the baker then owes you 20 cents. The time after that, you pay $2 again, increasing the debt of the baker to 60 cents. Finally, you get a loaf of bread for $1 and both you and the baker are debt-free again.

The moral of this story is that if you 'average out' the debt (in effect, the *rounding error*), you can approach any fractional value using only integers. We can use this same principle in the line-drawing function:

```
void draw_line(image_t *img,
               int x0, int y0, int x1, int y1,
               unsigned char value)
{
    double len = hypot(x1 - x0, y1 - y0);
    double dx = (x1 - x0)/len;
    double dy = (y1 - y0)/len;
    double x = x0;
    double y = y0;
    double err_x = 0;
    double err_y = 0;
    int i;

    for (i = 0; i < len; i++)
    {
        int ix = (int)(x + err_x + 0.5);
        int iy = (int)(y + err_y + 0.5);
        img->pixels[iy*img->width + ix] = value;

        if (fabs(dx) < fabs(dy))
            err_x += x - ix;
        else
            err_y += y - iy;

        x += dx;
        y += dy;
    }
}
```

The 'debt' is kept in the `err_x` and `err_y` variables. The `fabs(dx) < fabs(dy)` comparison is used so that the dithering is only performed in the direction more or less perpendicular to the line, otherwise 'gaps' can occur in the line. In every iteration, this debt (which can also be negative) is added to the current position before it is rounded, and the new value is determined.

When drawing lines with this version of the program, they may look 'noisy', but when viewed from sufficient distance (or at sufficiently high resolution) they look much 'smoother' than the original, aliased lines. Compare the lines in the figure below:

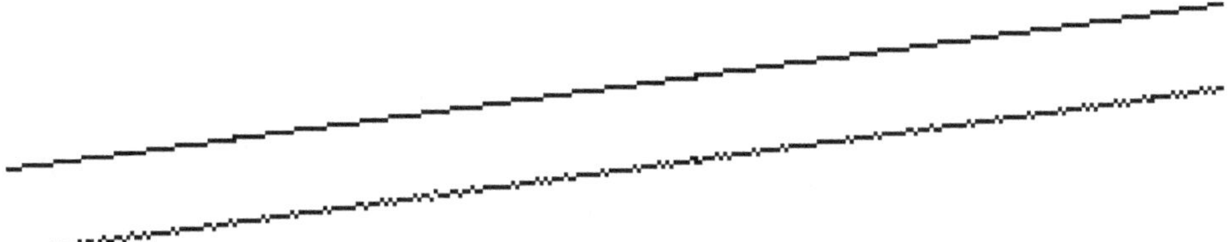

By holding the page at arms length and/or squinting a bit, you can see that the 'noisy' line will appear to be smoother.

This way of approaching fractions by dithering integers can be applied to other problems as well. A well-known example is that when a digital system (such as a micro-controller) is used to control an LED (light emitting diode) which it can switch either on or off, you can make the LED light up with any intensity in between by rapidly switching it on and off. Provided you do this with high enough frequency (say, several kilohertz) the human eye percieves the flickering as a continuous, dimmed light.

11.4.2 The Mandelbrot Set

We will start with a quick summary of *complex numbers*. Complex numbers are composed of a *real* part and an *imaginary* part. Imaginary numbers have the interesting property that their squares are negative. A complex number z can be written as $a + ib$, where a and b are both 'ordinary' real numbers and where i is defined so that $i^2 = -1$. When performing basic algebra on complex numbers, the following relations apply:

$$\begin{aligned}
(a + ib) + (c + id) &= (a + c) + i(b + d) \\
(a + ib) - (c + id) &= (a - c) + i(b - d) \\
(a + ib) \times (c + id) &= (ac - bd) + i(bc + ad) \\
(a + ib)/(c + id) &= (ac + bd)/(c^2 + d^2) + i(bc - ad)/(c^2 + d^2)
\end{aligned}$$

Now, consider a simple series of operations performed on a complex number c:

1. Let $z_0 = c$.
2. Set z_{n+1} to $z_n^2 + c$.
3. Repeat step 2 indefinitely.

It turns out there are two possible outcomes of this scheme: Depending on the initial value c, as $n \to \infty$, either z_n 'diverges' (*i.e.*, $|z_n| \to \infty$), or it stays within relatively small values forever. The set of values of c for which z_n does not diverge is called the *Mandelbrot set*, and it turns out that this set has some very interesting characteristics for which it has become very famous.

In this section, we will generate a graphical representation of this Mandelbrot set. Just like real numbers can be represented by points on a line, complex numbers can be represented by points in a 2D plane where the vertical axis is the imaginary axis.

The most straight-forward way of generating an image of the Mandelbrot set is to iterate over all the points in a rectangular image, set the value of c to the corresponding complex number, and determine whether or not this particular point belongs to the set. There is a problem though in that you cannot 'repeat indefinitely' because the program would hang at the first point encountered which is actually part of the set. Therefore, it is custom to change 'repeat indefinitely' into 'repeat a large number of times'. Also, it is not easy to figure out whether a value has 'diverged'. Fortunately, it can be shown that once $|z| > 2$, the value will diverge. This can therefore be used as an 'end criterion'.

The main body of the Mandelbrot-generating program will therefore be: For each pixel, set z_0 and c to the corresponding complex number, then loop a 'large number' of times setting $z_{n+1} = z_n^2 + c$, breaking off when the value of $|z_n| > 2$; if after the loop the value of $|z|$ is still below 2, the point is considered part of the Mandelbrot set and is colored black (for instance); otherwise, it is colored white.

In case you have a C99-compliant compiler, you can use the `_Complex` specifier (see page 47) and use complex numbers directly; in case your compiler doesn't support them, we'll define our own `complex_t` type along with the few functions we need:

```
typedef struct
{
    double re;
    double im;
} complex_t;

complex_t cplx_square(complex_t a)
{
    complex_t res;
    res.re = a.re*a.re - a.im*a.im;
    res.im = 2*a.im*a.re;
    return res;
}

complex_t cplx_add(complex_t a, complex_t b)
{
    complex_t res;
    res.re = a.re + b.re;
    res.im = a.im + b.im;
    return res;
}

double cplx_abs(complex_t a)
{
    return hypot(a.re, a.im);
}
```

The main program now is relatively easy. Two nested `for`-loops iterate over each pixel column and each pixel row; this example uses a 400×400 pixel grid. Then, these coordinates are scaled so they represent an area in the complex plane with real coordinates between -2 and 0.5 and imaginary coordinates between -1.25 and 1.25, because the interesting part of the image happens to fall between those coordinates. Next, 256 iterations are done for each coordinate and the pixel is set to the corresponding color value. Note that this version of the program makes the image a little bit more 'fancy' by using the exact value of `n` reached when the end criterion was met to assign an intermediate gray value, instead of simply coloring a pixel either black or white[2].

```
#include <stdio.h>
#include <stdlib.h>
#include <math.h>

/* Width and height of the resulting BMP image */
#define W 400
#define H 400

int main(void)
{
    color_t palette[256];
    image_t img;
    int x, y;

    FILE *fp = fopen("mandelbrot.bmp", "wb");
    if (!fp)
    {
        printf("Couldn't open mandelbrot.bmp for writing\n");
        return -1;
    }

    init_image(&img, W, H);
    for (y = 0; y < H; y++)
        for (x = 0; x < W; x++)
        {
            int n;
            /* Scale to -2 .. 0.5 in Re, -1.25 .. 1.25 in Im */
            complex_t c = { 2.5*x/W - 2.0, 2.5*y/H - 1.25 };
            complex_t z = c;
            for (n = 0; n < 255; n++)
            {
                z = cplx_add(cplx_square(z), c);
                if (cplx_abs(z) > 2)
                    break;
            }
            img.pixels[y*W + x] = n;
```

[2] Of course, this is for 'aesthetic purposes' only. A point is either part of the Mandelbrot set, or it isn't.

```
        }

    fill_palette(palette);
    write_BMP8(fp, img, palette);
    fclose(fp);
    deinit_image(&img);

    return 0;
}
```

Running this program should result in a file `mandelbrot.bmp` with an image looking somewhat like the figure below.

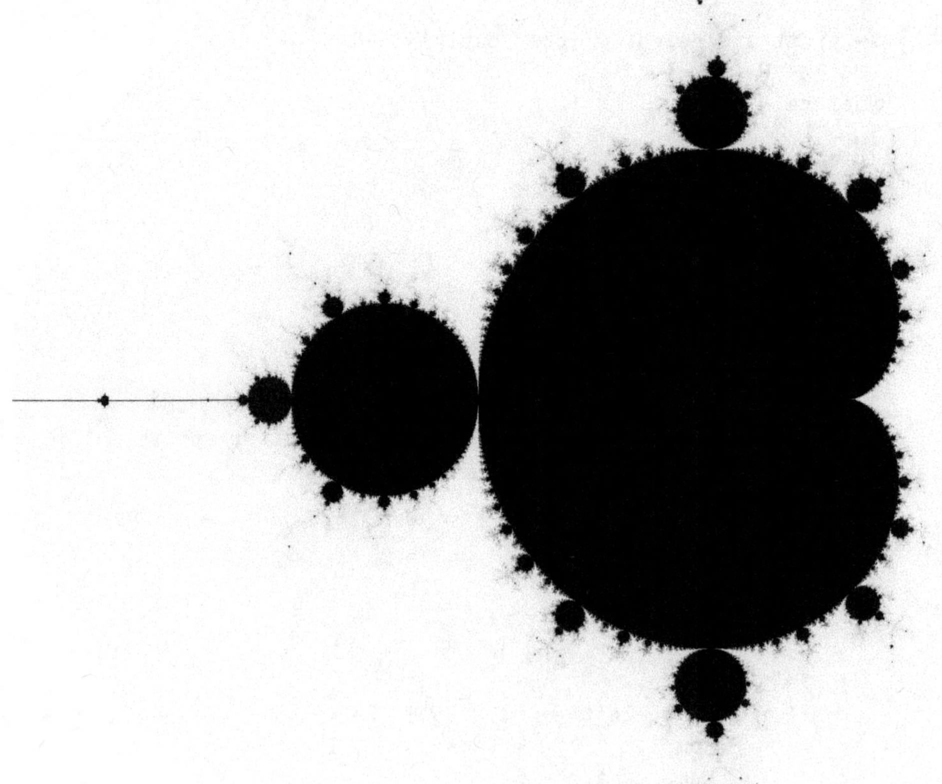

It can be rather mind-boggling that such a complex shape emerges from such a simple recipe. The boundary of this set is an example of a family of shapes called *fractals*. It has a fine structure at an arbitrary scale, it is too irregular to be described in traditional Euclidian geometric language but has a simple and recursive definition, and it is *self-similar:* zooming in on a part of them will reveal new shapes which resemble the original. Fractals are the subject of many a late-night science student conversation. Exercises 11.6 through 11.9 deal with this program and some of the interesting properties of the Mandelbrot set. An excellent read about fractals can be found in [3], by Mr. Benoit Mandelbrot himself.

11.5 Synopsis

Graphics programming is not standardized in C. There are, however, many libraries available for this purpose. Also, a simple 'trick' can be used where the the image contents are written to a file in the right format and another program is used to view the resulting image.

Graphics can be used for visualization (where graphics are *generated*, but you can also write programs which *operate on* images to enhance them or extract data from them. The latter is called *image processing*.

Computer graphics can be expressed in terms of *vector primitives* which are 'resolution independent', or as *bitmap graphics*, in which an image is viewed as a matrix of *pixels* which can each have their own color. The conversion of a vector representation to a bitmap is called *rendering*.

In programs with a graphical user interface (GUI), specific graphical primitives representing, for example, buttons are available (usually through the operating system). A computer screen itself presents a view on a bitmap as well (the *frame buffer*).

Since pixels are finite in size, most vectors cannot be rendered exactly. This can result in an effect called *aliasing*. Anti-aliasing using gray values is a remedy, as is *dithering* where the rounding error due to the discretization of the exact pixel positions is 'evened out' over the primitive to be rendered.

The *Mandelbrot set* is the most famous example of a *fractal*. It is quite easy to generate a graphical representation of this set; progams doing so are a grateful subject for optimization exercises.

An excellent (1200-page) reference on computer graphics is [1].

11.6 Other Languages

Although several graphical libraries (for generic vector drawing, bitmap manipulation, or even complex GUI building) are available for C, none of this is part of the language *per se*, nor is it a part of the standard library.

Some languages (such as C++) share this approach. Others either have simple graphics primitives (say, `SetPixel` or `DrawLine`) available in the language itself, such as many BASIC dialects, or have standardized libraries containing graphics and/or GUI support. Sometimes this is more or less tied to a certain operating system, sometimes it is 'cross-platform' (see also §12.4.1).

The latter can take two separate approaches. One is that a full set of graphics primitives and a complete GUI platform is available which is *independent* of the underlying platform. In essence, the operating system is only used to show the final rendered image on screen. A different approach is to 'wrap' the functionality already available on the platform; only the *way it is used* is standardized.

A drawback of the former method is that one can usually immediately spot that a given program is written in that particular language, since the program 'looks different' from other, 'native' programs running on that platform. This is often the case with programs written in Java, for instance.

A drawback of the latter is that you can only use a subset of the available functionality (*i.e.*, the 'lowest common denominator').

11.7 Questions and Exercises

11.1 Modify the BMP test code of §11.3.2 so the palette has more interesting colors than just gray values.

11.2 Use both the aliasing and dithering versions of the `draw_line()` function in §11.4.1 to draw lines at various slopes, and compare the results.

11.3 Using the original, aliasing version of `draw_line()`, draw a series of n lines starting in (0, 0), going to (399 - n, n), and continuing to (399, 399). The resulting image shows an interesting pattern called the *Moiré effect*. See what the image looks like when using the dithering version.

11.4 Implement a `draw_circle()` function.

11.5 If you take the Mandelbrot algorithm of §11.4.2 but use real numbers instead of complex numbers, which are the limits of the set? If you cannot find them analytically, write a program to provide numeric approximations.

11.6 If you look back to the section about optimizing your code (§8.3), you will find that it is generally a bad idea to make many function calls from within tight loops. Yet this is exactly what the Mandelbrot program on page 243 does. You can achieve quite a speedup (especially on older compilers which do not 'inline' functions) by not using the `square()` and `add()` functions, using separate variables for the real and imaginary parts, and typing out the full expressions for the calculations *inside* the loop. Do so, and measure the speedup.

11.7 Another (smaller) optimization is to simplify the end criterion, by noting that '$|z|^2 > 4$' is equivalent yet easier to calculate (because no square root is needed). See if that speeds up the program significantly.

11.8 $\star$ A major optimization in calculating a bitmap of the Mandelbrot set can be made by noticing that the set is *connected*, *i.e.*, it doesn't contain any isolated 'islands'. If you divide up the bitmap in subsequently smaller 'sub-rectangles' and only calculate the values of the pixels on the edges of these rectangles, then the entire rectangle is part of the set if all the edge pixels are part of the set, and vice versa (provided you did not pick too large a rectangle and the entire set fits within). Hence, the entire rectangle can be filled in without further calculations. Otherwise, the rectangle needs to be split up in smaller sub-rectangles which need to be investigated in turn. Devise a (recursive) implementation that uses this trick. Again, measure the speedup versus the original implementation.

11.9 One of the most striking features of fractals is that their level of detail is infinite: you can keep zooming in on an area of the fractal and new detail keeps emerging. Modify the Mandelbrot program so that it produces a 'zoomed in' version of the Mandelbrot set by taking a set of command line values to determine the scale (which was hard-coded to $[-2, 0.5]$ for the real part and $[-1.25, 1.25]$ for the imaginary part in the original program). You may have to increase the number of iterations performed for each coordinate (otherwise the level of detail will be bounded); don't forget to clip the resulting value so you don't specify colors > 255. Almost every sub-area around the perimeter of the set is interesting to zoom in on; if you look carefully you may find 'twin copies' of the original shape hidden in the small 'antennae' of the set.

11.10 ⋆ The 'self-similarity' of fractals can also be found in numerous examples in nature. One such example is a fern leaf (as shown on the cover of this book). Using the `draw_line()` function from §11.4.1 and a recursive algorithm, see if you can produce a graphic like the one below.

The basic structure of this fern leaf is a line which forks into two smaller lines. At the end of each of these lines, there is again a fork, etcetera. As another hint, here's the signature of the function which was used in the recursion to generate this image:

```
void fern(image_t *img, double x, double y,
          double angle, double length, int leftright);
```

Note that the lines are characterized by their starting position (x, y), a direction, and a length. In each recursion, the direction and length were modified slightly for the 'main branch' (0.05 radians and a factor of 0.92, respectively) and by a full radian and a factor of ± 0.4 for the 'leaves'. The sign was flipped each recursion to get a nice left/right distribution of the leaves, for which the `leftright` toggle was used. Prevent 'infinite recursion' by immediately returning once `length` is below a certain limit (of, say, a few pixels).

Chapter 12
Libraries

I have always imagined that Paradise will be a kind of library.

Jorge Luis Borges

In §10.5, the concept of *libraries* was explained. This chapter focuses entirely on libraries, with an explanation of how they work followed by a brief overview of some existing libraries available to C programmers. Brief, because a thorough overview would be a book all in itself. In fact, for many of the libraries below there exist several dedicated books.

12.1 Library Mechanics

12.1.1 Static and Shared Libraries

As explained in §10.5, a *library* is basically a pre-compiled collection of functions. In its simplest form, you could just keep one single pre-compiled object file around and include it when linking your own program. Once a library grows, it could be beneficial to split it up into several source files (for reasons explained in chapter 10). It is a nuisance to have to keep a *bunch* of object files around and link them into your program each time. Fortunately, most linkers allow grouping those object files into a single file. Usually, such a 'grouped file' is what people mean when they refer to a *library*. On UNIX systems, such a library is usually given a `.a` extension (for archive); on Windows, it is `.LIB`.

In fact, the only difference between a program and a library, to the linker, is whether the 'glue code' (for starting the program on this particular operating system) is linked in. For a library, there doesn't need to be a `main()` function, as this is provided by any program *using* the library.

Sometimes the 'ordinary' linker can be instructed to link a set of object files into a library, sometimes specific tools are used. As an example, on most UNIX systems (including Linux and Mac OS X) there is the `libtool` utility. Suppose you want to create a library on such a system, you would use the command

`libtool -o mylibrary.a` *list of object files*

You can then use this library in other programs by adding it to the list of object files to link:

```
cc -o myprogram list of myprogram's object files mylibrary.a
```

Note that you may have to specify the library last, since some linkers will only look for symbols undefined *so far* (with the benefit of preventing the linking in of code which isn't used by the program anyway).

Using Visual Studio on Windows, it is a setting of your project (projects were explained in chapter 10). In fact, there is usually a predefined project type for shared libraries, which is available from the same place as your usual 'Win32 console application' type.

Libraries like these are called *static* libraries. You have to link them in to each program which uses them, and if you want to change the library (for example by implementing a more efficient version of a certain function) you will have to re-link all applications using it.

There is another type of library, which is called a *dynamic* or *shared* library. Most operating systems support this type of library, which resides on your system in one specific place, and which is linked in 'by reference' into all programs using it. When the operating system is asked to run a program which uses this type of libraries, it will look for these libraries and load them into memory too. In an analogy, you could say that the program only contains a 'pointer' to the library.

This has two advantages: One is that each program has a smaller executable size than if the corresponding functions were linked into the program directly, and that the code of the library is present on your system only once. A second, very interesting advantage is that the library can be changed *after* linking your program. As long as the new library has the same name, the operating system will load the new library into memory when it is asked to run your program. When the library is improved, all programs linking dynamically to this library will automatically use the updated functionality!

On Windows systems, this type of library is called a `DLL` (for **d**ynamically **l**oaded **l**ibrary); on UNIX systems they usually have a `.so` extension (for **s**hared **o**bject). Making shared libraries yourself is beyond the scope of this book.

12.1.2 Installing Libraries

Some of the libraries covered in this chapter are pre-installed on your platform (because they came with the operating system or with your compiler); others may have been installed by your system administrator, and if you are working on your own PC, you will have to install them yourself.

Since shared libraries will have to be 'found' by the operating system when they are referenced by a program about to be run, they will have to be placed in a location where the operating system can find them. On Windows, for example, it is sufficient to put a `DLL` in the same directory as your executable. On many UNIX systems, it has to be in a specific directory (`/usr/local/lib` for example), or the directory containing the library will have to be added to the 'search path' for libraries. If you are installing a third party library, its installer will most likely take care of these details. On UNIX systems, making a library available is usually done by unpacking its source code (a large number of libraries are distributed in source code archives), followed by

```
./configure && make && make install
```

which will configure the library for your particular system, build it, and install it (and its associated header files). After this, you can use it in your own programs by including the header files it provides and linking against the library. On Windows systems, you specify which shared libraries to link to by listing them in your project settings, or with the `/link` command line switch. On UNIX systems, you specify the library with the `-l` switch. By convention, if a certain library is called `libsomething.so`, you specify

```
cc -o myprogram -lsomething myprogram.c
```

(In other words, the '`lib`' part of `libsomething.so` is implicit.) The library will have to be installed in a place where the linker can find it, otherwise you will get an error. If the header files for this particular library have been installed somewhere the compiler can't automatically find them, remember from §10.4.4 that you can specify their location using the `-I` (UNIX systems) or `/I` (Visual Studio) switches.

12.1.3 Versioning

The fact that shared libraries can be updated independently of the programs linking to them was mentioned as an advantage above, but it is also a drawback. It means that when a library is updated, its developer will have to make sure that the new version is *compatible* with the previous one. Usually, simply *adding* new functions is not a problem. Changing existing functions (even to 'improve' them) needs to be done with caution, because existing programs may depend on the current functionality. If the signature of a function is changed (taking an extra parameter, for example), existing programs may even *crash* after the library is updated.

The problem is also that installing a program with its associated libraries may put *older* versions of certain libraries on your system, overwriting the newer ones which other programs depend on. In other words, installing a program on your system may break other programs already present!

This problem is often referred to as '`DLL` hell'. The solution is to use a 'versioning scheme', in which you give each revision of a library a certain version number. Often, libraries specify a 'major version' and a 'minor version' number. The agreement is usually that libraries having the same major version are compatible with each other, and that the minor versions are *backwards compatible*, meaning that you can always safely use a newer version (but not necessarily an older one). Sometimes there is a third version number (usually called a 'patch level') which is used to indicate small bug fixes or optimizations.

The idea is then that an installer of a certain program checks whether there is already a version of its libraries available on the system which has the same major version number, and if so, only installs its own version if that has a larger minor version number than the one already present.

As an example, suppose that a program is installed which needs at least version 1.5 of a certain library. On a system where versions 1.4 and 2.0 are already present, it would install version 1.5. Version 2.0 may have a completely different set of functions, being a major revision, and version 1.4 is missing some of the functionality in 1.5 that it needs. If there are other programs installed on the system which need version 1.4 of the library, these can safely use 1.5 because this is 'backwards compatible'.

On a system with only version 1.7 of this particular library available, it would *not* install its own version 1.5, because it can safely assume that version 1.7 offers all the functionality it needs.

12.2 A Word on Licenses

It is important to know what *license* a certain library is provided under, especially if you consider distributing your software. When you write a program using a certain library only for your own use (as is quite often the case with scientific programs: You are only interested in the *results* of the program), it is often quite clear: You simply buy the library (if it's commercial) or download it (if it's free).

It's different when you distribute your software to others, even if you do that non-commercially (*i.e.*, if you don't *sell* your program). After all, you will have to distribute the library as well; otherwise your program won't run. Different vendors have different requirements in that case.

Sometimes, you will have to buy a new license for the library for each copy you distribute. In other cases there are special 'redistributables'; usually only the actual library itself, without the accompanying header files and documentation. These can either be distributed freely or at a reduced price.

A special case is when the library is under the GNU Public License (GPL), which is a very popular Open Source license. Open Source, as the name implies, means that the source code for the program should be available to anyone. The idea is that if everyone opens up their source code, people can study other people's work, learn from it, and make improvements. Some Open Source licenses are very 'lax', and boil down to 'You are free to do what you want with this code, as long as you don't pretend you wrote it, and make a note in your documentation that you use it'.

The drawback of the 'lenient' licenses is that commercial entities can profit from the work of the library developers, yet the developer community gets nothing in return. To prevent this 'one-way sharing', the GNU Public License was constructed. GNU is a recursive acronym which stands for 'GNU's Not UNIX', and it set out to provide a free, open source replacement for the UNIX operating system. Linux, although not written by the people behind GNU, is released under the GPL, and a lot of programs running on top of the Linux kernel in most distributions of it are GNU utilities. The excellent GCC (GNU Compiler Collection) is an example of this. You often see the combination referred to as 'GNU/Linux'.

What the GPL states is that you are free[1] to base your work on code which is under the GPL, but that when *distributing* your work, you have to provide the source code as well, and it automatically has to fall under the GPL too. This is sometimes called the 'viral nature' of the GPL. The idea is that modifications and improvements to GPL software are automatically shared with the developer community, and that the body of code under the GPL automatically grows whenever someone makes use of GPL code in his/her own program.

Libraries, however, form a 'grey area'. If you only *link* to a library, is that 'basing your work' on it? Especially if you are linking to it dynamically (see §12.1.1), your code and the code from the library are combined only when it is launched; your program does not really *contain* any GPL code.

There is a special version of the GPL called the LGPL (originally for 'Library GPL' but now often used as 'Lesser GPL') which states that you are free to use a library released under the LGPL, provided that any modifications you make *to the library itself* must be contributed back to the community, but the program *using* the library does not automatically become LGPL'd as well.

Finally, it may be interesting to note that many library vendors offer special 'academic discounts' for schools and universities, or offer 'dual licenses'. For example, a library may be under the GPL but you can buy an exception to the library from its copyright holders – this way, if you cannot or don't want to open up your source code, at least the developers get a financial reward which can be used to fund further development.

12.3 The Standard C Library

After all this talk about *how* to use libraries, let's take a look at *what* you can use. The first library we'll look at is bound to be present on your system (because it came with the C compiler), and it's called the *standard* library. We've actually been using many functions from this library already, and since it's a part of C, we did not have to specify any special compiler flags to do so. This section will give a brief overview of the functionality available in the standard library. The C manual ([2]) contains a full overview of the standard library in more detail, but no C99 version of this book is planned. The ISO standard (ISO/IEC 9899:1999) contains a full chapter about the C99 standard library.

There are usually no restrictions on distributing programs which use standard library code. Most computer systems already have the necessary shared libraries installed anyway.

12.3.1 Math

In §3.9, much of the available functionality for doing mathematics was already mentioned. On some systems you may have to link to the 'math' part of the standard library

[1] To distinguish GPL software from other 'free' software, they often use the motto 'Free as in *free speech*, not just free as in *free beer*'.

explicitly by specifying '`-lm`' on the command line, but on most systems everything in the standard library is automatically found by the linker.

By including `math.h`, you get access to the trigonometric functions `sin`, `cos`, `tan`, `asin`, `acos`, `atan`, and `atan2`, as well as the hyperbolic functions `sinh`, `cosh`, `tanh`, `asinh`, `acosh`, and `atanh`. There are also `exp` and `log` (remember that `log(x)` actually computes what mathematicians often call '$\ln x$', *i.e.*, the *natural* (base-e) logarithm). For $\log_{10} x$ and $\log_2 x$, there are `log10(x)` and `log2(x)`, respectively.

There are also the functions `pow(x,y)` for x^y, `sqrt(x)` for $\sqrt{x}$, and (mandated by C99 but also available in many earlier implementations) `cbrt(x)` for the cube root $\sqrt[3]{x}$ and `hypot(x,y)` for $\sqrt{x^2+y^2}$. The absolute value of a floating point value, $|x|$, is given by `fabs(x)`. The integer version of the latter (`abs(x)`) is in the header `stdlib.h`.

By including `complex.h` (on a C99-compliant system), you also have complex versions of all of the above, prefixed with a `c` (with the exceptions of `cbrt` and `hypot`, of which there are no complex versions in the C99 standard).

There are also more 'exotic' functions in `math.h` such as `expm1(x)`, which computes $e^x - 1$, and `log1p(x)`, which computes $\ln(1+x)$. For small values of `x`, these functions are usually more accurate than using `exp(x) - 1` and `log(1 + x)`. This is because of the nature of floating point numbers, explained in §1.3.2.

There are also functions to compute the 'error function' given by

$$\operatorname{erf} x = \frac{2}{\sqrt{\pi}} \int_0^x e^{-t^2}\, \mathrm{d}t$$

and the Gamma function $\Gamma(x)$ (a sort of 'generalized factorial'), namely `erf(x)` and `tgamma(x)`, respectively.

Some functions for rounding floating point values are `ceil(x)` for $\lceil x \rceil$ (the smallest integer value not less than x), `floor(x)` for $\lfloor x \rfloor$ (the largest integer value not greater than x), and `round(x)`, which rounds to the nearest integer value, rounding halfway cases away from zero.

There are also some 'classification macros' which you can use to determine whether a floating point variable is representing one of the 'special numbers' mentioned in section 3.9, including

```
int isfinite(x);
int isinf(x);
int isnan(x);
```

Remember that C thinks that $\frac{1}{0} = \infty$ and $\frac{-1}{0} = -\infty$, where both expressions arguably should yield NaN (see exercise 3.7).

The C99 standard contains several more mathematical functions, and many compilers include functions which are not part of the standard (such as Bessel functions). Before using separate libraries for special functions, it could be worthwhile to check whether your 'standard' library already includes them. Of course, this then limits your ability to 'port' your program to a different platform.

12.3.2 Input/Output

The functionality in the standard library for input and output is available after including `stdio.h`. Dealing with files was introducted in chapter 7, but the standard library offers some more file handling functions, such as

```
int remove(const char *filename);
```

which deletes the given file (a zero return value means the file was successfully removed), and

```
int rename(const char *oldname, const char *newname);
```

which renames a the file `oldname` to `newname`.

The functions `printf` (§3.5) and `scanf` (§4.4) are also in `stdio.h`, as well as their more 'generic' versions `fprintf` and `fscanf`. The function

```
int snprintf(char *buf, size_t n, const char *fmt, ...);
```

is like `printf`, but the resulting output is written into the string `buf` instead of to `stdout`. The output will be at most `n` characters, so you should pass the size of the string pointed to by `buf` there to prevent a 'buffer overflow' (see page 102) from occurring. There *is* a function `sprintf` which does not take this extra 'size' parameter, but because this is inherently 'unsafe' you should never use it.

There is a matching

```
int sscanf(const char *buf, const char *fmt, ...);
```

which is like `scanf`, but reads the input from the string `buf` instead of from `stdin`.

There are also some functions for one-character-at-a-time input/output, such as

```
int fgetc(FILE *fp);
```

which reads a single character from `fp`. The reason the return value is an `int` instead of a `char` is that it can also return `EOF` when there are no more characters available. There is a corresponding

```
int fputc(int c, FILE *fp);
```

which *writes* a single character `c`.

You would expect that the 'shortcut' versions dealing with `stdout` and `stdin` are called `getc` and `putc`, but they are

```
int getchar(void);
int putchar(int c);
```

instead. The functions `getc` and `putc` do exist, but they are equivalent to `fgetc` and `fputc` (only the standard allows for them to be implemented as macros).

There are convenient functions for reading and writing an entire string at once, namely

```
char *fgets(char *s, int n, FILE *fp);
int fputs(const char *s, FILE *fp);
```

The first one reads at most one character less than the number `n` into the string `s` (like in `snprintf`, you can simply fill in the size of the string `s` here to prevent buffer overflows). A null character is written immediately after the last character read into the array, so the resulting string is zero-terminated. The `fgets` function returns `s` if successful. Otherwise, a null pointer is returned. No additional characters are read after a newline character (which is retained) or after end-of-file.

The `fputs` function writes the string `s` to `fp`, including any newline characters in the string, but excluding its terminating zero.

There is a `puts` function which always uses `stdout`, but the corresponding `gets` function (reading from `stdin`) doesn't allow you to specify the buffer size and should therefore be avoided.

12.3.3 String Handling

The header `string.h` contains useful functions for dealing with strings, such as

```
size_t strlen(const char *s);
```

which returns the length of the string pointed to by `s`; that is, the number of characters excluding the terminating null character.

```
char *strncpy(char *dst, const char *src, size_t n);
```

This copies (at most `n` characters of) the string `src` to the array pointed to by `dst`. The function returns `dst`. By passing the size of `dst` as `n`, you can circumvent buffer overflows, but note that if there is no null characters in the first `n` characters of `src`, the result will not be null-terminated! To be safe, you should assign `dst[n - 1] = 0` afterwards.
Note that there is also a `strcpy`, which doesn't take the `size_t` parameter, and thus should be avoided.

```
char *strncat(char *s1, const char *s2, size_t n);
```

This function appends (at most `n` characters of) string `s2` to `s1`. This means that you must pass the *remaining* space in `s1` as `n`, not the *total* room available. Contrary to `strncpy`, `strncat` *does* append a terminating null character. The maximum number of characters that can end up in the array pointed to by `s1` is therefore `strlen(s1) + n + 1`.

There is also a `strcat` not taking the `size_t` parameter – again, don't use it.

Some functions for comparing strings are

```
int strcmp(const char *s1, const char *s2);
```

which compares the strings pointed to by s1 and s2, returning zero if they are equal, less than zero if s1 is alphabetically before s2, and a value greater than zero if s1 is alphabetically after s2, and

```
int strncmp(const char *s1, const char *s2, size_t n);
```

which compares at most the first n characters. Note that unlike the functions *copying* characters mentioned above, it is entirely safe to use strcmp instead of strncmp.

Some platforms also offer strcasecmp (or sometimes stricmp) which allows a 'case-insensitive' string comparison.

```
char *strchr(const char *s, int c);
```

returns a pointer to the first occurrence of c in the string pointed to by s. The terminating null character is considered to be part of the string (*i.e.*, you can pass zero for c). If c is not part of the string, strchr will return a null pointer.

```
char *strrchr(const char *s, int c);
```

is like strchr, but it returns a pointer to the *last* occurrence of c in s.

```
char *strstr(const char *s1, const char *s2);
```

locates the first occurrence of s2 in s1 and returns a pointer to the located string. If the string s2 is not found in s1, or s2 is an empty string, the function returns s1.

There are several more functions related to string handling, including functions to 'tokenize' a string (strtok), or to compute the length of the maximum initial segment of the string pointed to by s1 which consists entirely of characters *not* from the string pointed to by s2 (there really is a function which does that: strcspn), but these are rather rarely used in 'scientific' programs.

The string.h header also contains some 'generic' memory manipulation functions:

```
void *memcpy(void *dst, const void *src, size_t n);
```

copies n bytes from src to dst, and returns dst. The memory must not 'overlap' (*i.e.*, src is within n bytes from dst).

```
void *memmove(void *dst, const void *src, size_t n);
```

is like memcpy, but this version can handle 'overlapping' memory regions.

```
void *memset(void *p, int c, size_t n);
```

sets n bytes starting from p to c. This is usually used to 'clear' a certain block of memory, for instance a struct:

```
struct s;
/* ... */
memset(&s, 0, sizeof(s));
```

In the header ctype.h there are some character classification functions, which take an int (so that they can also be passed an EOF value), and return an int which is nonzero ('true') if the argument is in the specified class of characters:

int f(int c)	character class of c
islower	a lowercase letter (a – z)
isupper	an uppercase letter (A – Z)
isalpha	a letter (islower or isupper is true)
isdigit	a digit (0 – 9)
isalnum	an alphanumeric character (isalpha or isdigit is true)
isgraph	any printable character excluding space
isprint	any printable character including space
isspace	a 'white-space' character (space, tab, new-line, etc.)

There are a few more, but these are the most common. There are also character case conversion functions:

```
int tolower(int c);
```

converts an uppercase letter to a corresponding lowercase letter. If you pass in a character for which isupper is false, the argument is returned unchanged.

```
int toupper(int c);
```

converts the other way around: a lowercase letter is converted into its corresponding uppercase letter.

C99 also includes 'locale support', so that functions like the above work in non-english languages as well. For example, in the standard locale, characters like ü and Ç are unknown.

A convenient function in string.h is

```
char *strerror(int errnum);
```

which maps the number errnum to a message string. In §3.10, it was suggested to use 'meaningful #defines' for error code return values. The C standard library does something similar. In most cases, the standard library sets a special global int variable errno when something goes wrong during the execution of a function. For example, the fopen function returns a null pointer when it can't open the given file, but sets errno to indicate the *reason* (for example, the file doesn't exist at all, or you have insufficient permissions to open the file). The error constants are available by including errno.h, but using strerror you can convert a still-rather-cryptic error code such as ENOENT into a human-readable message like 'No such file or directory'. You could use this in your programs like so:

```
FILE *fp = fopen(argv[1], "r");
if (!fp)
{
    fprintf(stderr, "Couldn't open %s: %s\n", argv[1],
            strerror(errno));
    return 1;
}
```

Since the above 'code pattern' occurs often, there is an even more convenient function for this, available in stdio.h:

```
void perror(const char *msg);
```

which is equivalent to

```
fprintf(stderr, "%s: %s\n", msg, strerror(errno));
```

(if `msg` is a null pointer, it will only print out `strerror(errno)`).

12.3.4 Date and Time

The header `time.h` contains definitions for manipulating time values.

```
clock_t clock(void);
```

returns (an approximation to) the 'processor time' used by the program. The `clock_t` type is an arithmetic type capable of representing times, and to convert this to seconds you can divide it by the value of the macro `CLOCKS_PER_SEC`, which is also defined in `time.h`.

The function

```
time_t time(time_t *t);
```

returns the (best approximation of the platform to) the current calendar time. If you pass it a pointer to a `time_t` value, then the return value is also assigned to the value pointed to. You can also pass `NULL`, though. The `time_t` type (like `clock_t`) is defined as an arithmetic type capable of representing times, but the encoding of the value is unspecified (it is usually something like 'seconds passed since January 1st, 1970').

There is another type defined in `time.h`, which is `struct tm`, with the fields

```
int tm_sec;   // seconds after the minute [0, 60]
int tm_min;   // minutes after the hour [0, 59]
int tm_hour;  // hours since midnight [0, 23]
int tm_mday;  // day of the month [1, 31]
int tm_mon;   // months since January [0, 11]
int tm_year;  // years since 1900
int tm_wday;  // days since Sunday [0, 6]
int tm_yday;  // days since January 1 [0, 365]
int tm_isdst; // Daylight Saving Time flag
```

The range $[0, 60]$ for `tm_sec` is to allow a positive leap second. The `tm_isdst` field is positive if Daylight Saving Time is in effect, zero if it isn't, and negative if this information is unavailable.

There are several functions for converting a `time_t` value (which is easier to perform calculations with) to a `struct tm` (which is easier to interpret and construct).

```
time_t mktime(struct tm *t);
```

converts the time structure pointed to by `t`, expressed as local time, to a `time_t` value, or to -1 if it cannot be represented. The original values of the `tm_wday` and `tm_yday` fields are ignored in the conversion, but are set to their appropriate values on succesful completion (this is why `t` is not a `const` pointer).

Conversion 'the other way around' is done with

```
struct tm *gmtime(const time_t *t);
struct tm *localtime(const time_t *t);
```

Given a pointer to a `time_t` value, they return a pointer to a broken-down time, either expressed as UTC or as local time, respectively. If the specified time cannot be converted to the corresponding broken-down time structure, a null pointer is returned.

```
char *asctime(const struct tm *t);
```

This function converts the broken-down time pointed to by `t` into a string of the form

```
Day Mon dd hh:mm:ss yyyy\n
```

where `Day` is a three-letter abbreviation of the day of the week, and `Mon` a three-letter abbreviation of the month.

The shortcut

```
char *ctime(const time_t *t);
```

is equivalent to `asctime(localtime(t))`. There is also a more generic `strftime()` function which takes a format string (similar to, and just as extensive as, the one used in `printf()`) which allows very precise formatting of a `struct tm` into a character string. Even a short description of each of the available format characters takes three pages and is outside the scope of this book.

12.3.5 General Utilities

The header `limits.h` defines a number of macros which give information about certain types, depending on your platform and implementation, such as `INT_MAX` (the maximum value which can be represented with a (signed) `int`), `INT_MIN` (its minimum, *i.e.,* most negative value), `UINT_MAX` (the maximum value representable with an `unsigned int`). There are also `SHRT_MIN`, `SHRT_MAX`, and `USHRT_MAX` for the limits of the `short` data type, and `LONG_MIN`, `LONG_MAX`, and `ULONG_MAX` for the `long`.

Note that the C language doesn't really specify what the exact number of bits in a certain data type is, only that an `int` can hold values *at least* as large as a `short` can, and a `long` can hold values *at least* as large as an `int` can. It is therefore perfectly acceptable if `short`s, `int`s, and `long`s are all 32-bit integer data types. In many implementations though, a `short` is 16 bits.

C99 adds the `stdint.h` header file which has (much) more detailed information about types, such as 'integer types having certain exact widths' (*e.g.,* `int32_t` for an integer type with exactly 32 bits), 'integer types with *at least* certain specified widths' (*e.g.*, uint_least16_t for an unsigned integer type with 16 bits of width or more), 'an integer type with a certain minimum width, which is the the fastest on a certain platform to operate with' (*e.g.*, int_fast16_t for the fastest integer type with at least 16 bits, which may well be a 32 bit integer in reality).

The stdlib.h header file contains some memory management functions:

```
void *malloc(size_t size);
```

allocates space for an object whose size is specified by size and whose value is indeterminate (*i.e.*, the contents of the memory are unspecified). The function returns a pointer to the allocated space, or a null pointer if the allocation fails (because the system is out of memory).

```
void *calloc(size_t nmemb, size_t size);
```

is like malloc(); it allocates space for (and returns a pointer to) an array of nmemb objects, each of whose size is size. Unlike with malloc(), the memory is initialized to all bits zero.

```
void *realloc(void *ptr, size_t newsize);
```

deallocates the old memory area pointed to by ptr and returns a pointer to a new area of size newsize. It will make sure that the new memory contains the same data as the old one, up to the lesser of its original size and newsize. The ptr passed to realloc() must be the result of an earlier malloc(), calloc(), or realloc() call. Otherwise, the result is unspecified. If the new memory cannot be allocated, the old memory is not deallocated and its value remains unchanged. In this case, a null pointer is returned.

```
void free(void *ptr);
```

deallocates the space pointed to by ptr. This pointer must be the result of an earlier call to malloc(), calloc(), or realloc().

There is also a function which lets you use the operating system to perform a certain command as if you typed it on its command line. Of course, this is non-portable (*i.e.*, on a UNIX or Mac OS X system, a different command syntax is used than on a Windows system).

```
int system(const char *cmdline);
```

If cmdline is a null pointer, the system() function checks whether a command processor is available (and returns nonzero only if so). Otherwise, the string cmdline is passed to the available command processor and is executed. The return value of the system() call is implementation-defined (usually it is the return value of the executed program or command). An example would be system("copy a.txt b.txt") on Windows, which would copy the file a.txt to b.txt. On a UNIX or Mac OS X system, the corresponding call would be system("cp a.txt b.txt").

There are also sorting and searching functions in stdlib.h. The function

```
void qsort(const void *base, size_t nmemb, size_t size,
        int (*compar)(const void *, const void *));
```

sorts an array base of nmemb objects, each of which is size bytes. The contents of the array are sorted in ascending order according to a 'comparison function' which is passed as the compar argument. This comparison function is called with two pointers to the objects being compared. It must return an integer value of zero if those objects are identical, less than zero if the first is 'less than' the second, and greater than zero if the first is 'greater than' the second. The name qsort implies that the algorithm used is Hoare's *quicksort* (and it usually is), but the standard does not actually specify *how* qsort sorts the array, nor does it make any performance guarantees.

```
void *bsearch(const void *key, const void *base,
            size_t nmemb, size_t size,
            int (*compar)(const void *, const void *));
```

searches an array base of nmemb objects, each of which is size bytes, for an element that matches the object pointed to by key. The compar function has similar requirements as the one for qsort(), above.

The function returns a pointer to a matching element in the array, or a null pointer if none is found. The array must be *partially sorted:* all elements less than key need to preceed it in the array, and all elements greater than key must come after it. In practise, the array is simply sorted in ascending order.

Although the standard does not specify *how* bsearch finds the element matching key, in practise this is done by performing a 'binary search' (hence the name), much like the algorithm described on page 144.

12.4 Using the Operating System

As was explained in §1.4, the main task of the operating system is 'virtualizing' the machine your program runs on. The operating system tries to schedule different programs to use the CPU most effectively, takes care of launching new programs when requested by the user, makes sure one errant program cannot disturb other running programs, etc.

However, this 'basic' task of the operating system has gradually expanded. When talking about 'the operating system', one usually no longer means just the 'kernel' (the part which takes care of the tasks mentioned above), but also the growing set of libraries and services provided with it, such as a windowing system, support for advanced graphical operations, all kinds of media formats, playback of DVDs, etc.

Much of this functionality is available to programs running on a particular operating system by calling functions in libraries provided with the system. The drawback of using these libraries is that your program is no longer *portable*, *i.e.*, it is tied to that particular operating system.

Most systems provide access to their functionality in the form of a C library, although this may not necessarily be the recommended way. For instance, on Mac OS X, you can program exclusively in C using the 'Carbon' framework, but it is easier to use the 'Cocoa' framework which is programmed in Objective-C (an object-oriented language based on C).

12.4.1 Cross-Platform Development

There are cross-platform development libraries available (usually not from operating system vendors, as you can imagine) which allow you to make programs that run on several platforms, even using graphical user interfaces. Sometimes, such libraries simply provide 'wrappers' for the existing functionality offered by the operating system. In that case, the library calls platform-specific functionality so that when moving an application to a different platform (porting), it is only the *library* which needs to be converted to use the specific functionality of the new platform, and the application itself can remain unchanged.

A simple example is the 'curses' or 'ncurses' library, which provides a standard interface to functionality available in computer terminals to reposition the cursor anywhere on the screen, so that you don't need to redraw the entire screen everytime you only change a few characters on it. You can imagine that this makes developing, say, a text editor much easier.

It can be worthwhile to look for cross-platform libraries even for functionality which your current operating system provides natively. The experience gained with such libraries is useful if you later move to a different platform, and it makes your program more portable. On the other hand, if you are planning on distributing your program only to people using the same platform, using operating system provided functionality means you will have to distribute fewer libraries with your program.

As mentioned in §2.5, there are also languages which take the 'cross-platform' idiom a step further, such as Java. This language is not compiled to native machine code of the current platform, but to a special 'byte-code' which is executed by a virtual machine. As long as this virtual machine is available on a (real) platform, all Java programs should run on it. More recently, the .NET languages (such as C#) took a similar approach, and although they are a Microsoft invention, there is a Linux implementation of the corresponding virtual machine (developed in the 'Mono' project). In principle, C could be compiled to such a 'virtual machine' as well. But since there are C compilers for just about any platform imaginable, there is little research in this area.

12.5 Numerical Libraries

Many scientific problems can be attacked with the help of computers. As was shown in Chapter 9, sometimes you can get interesting results with relatively simple programs. On the other hand, sometimes there is a lot of number-crunching involved.

Often, a 'big' problem can be reduced to (a series of) 'known' problems, which have been investigated separately. When there are numerical approximations or other computational approaches available, the corresponding algorithms are usually found in literature. And when those algorithms find widespread use (as a surprisingly large number of algorithms do), chances are that they are available in a library somewhere.

There is a wealth of libraries available offering many numerical algorithms, from calculating special functions to inverting large sparse matrices. Some libraries focus entirely on efficiently solving one particular type of problems (such as calculating discrete Fourier transforms), whereas others are a collection of dozens (sometimes hundreds) of loosely related algorithms.

We will highlight at a small selection of numerical libraries in this section.

12.5.1 BLAS and LAPACK

Sometimes, libraries make use of other libraries in turn. A good example is BLAS. The BLAS (Basic Linear Algebra Subprograms) are routines that provide standard building blocks for performing basic vector and matrix operations, like the functions for inner and outer products described in chapters 5 and 6. The 'Level 1 BLAS' perform scalar, vector, and vector-vector operations, the 'Level 2 BLAS' perform matrix-vector operations, and the 'Level 3 BLAS' perform matrix-matrix operations. They were originally developed in FORTRAN (like most computing intensive software), but there is a C interface to the routines as well. Visit `http://www.netlib.org/blas` for more information.

Another common library is LAPACK, which stands for 'Linear Algebra Package'. It was also originally written in FORTRAN, but interfaces to C are available (search for 'CLAPACK'). The LAPACK routines are written so that as much as possible of the computation is performed by calls to the BLAS.

Since these routines are well standardized, several systems vendors provide libraries specifically optimized for their computer system (or processor). For example, Intel has the Math Kernel Library (MKL) for their processors, and AMD has the ACML. You can also look for ATLAS (Automatically Tuned Linear Algebra Software), which is an ongoing research effort applying empirical techniques in order to provide *portable* performance.

12.5.2 Numerical Recipes

Numerical Recipes is originally a text book containing explanations of more than 300 numerical algorithms, covering linear equations, matrices and arrays, curves, splines, polynomials, functions, roots, series, integrals, eigenvectors, FFT and other transforms, distributions, statistics, ordinary and partial differential equations, and more. Each algorithm was accompanied by an implementation (in FORTRAN), which made the book very popular since it could be used as a text book to learn numerical methods from, but also offered the 'shortcut' of simply browsing for the needed algorithm and typing in the corresponding code.

The book was later 'translated' to other languages than FORTRAN, and the C edition ('Numerical Recipes in C, 2nd edition', [4]) had nearly 1000 pages. This edition of the book has been 'obsoleted' by a C++version, and is even available online (including all source code) on `http://www.nrbook.com`. The material is still extremely worthwhile, although a (common) point of critique is that the FORTRAN 'heritage' shows in the code (with array indices starting at 1 instead of 0, for example).

The main site for Numerical Recipes is `http://www.nr.com`. The algorithms found in the book can be used without restriction (*i.e.*, you don't need a special license to use them in your programs).

12.5.3 NAG

The Numerical Algorithms Group sells a library of high-quality and efficient numerical algorithms. `http://www.nag.com`.

The implementations in this library also have a FORTRAN heritage. Originally in FORTRAN, names had to be unique in their first 6 characters (the rest was ignored), so the library functions have rather 'cryptic' names, such as `c06ebc()`. Luckily, there are also 'long names' available, which are more meaningful, such as `nag_fft_hermitian()`.

One thing to note is that programs using the NAG library often use NAG-defined data types to replace and augment the C provided types (for example, `Integer` which is normally defined as `long`, or `Pointer` which is a `void *`). Especially when a program is also using the 'short names' for functions, it can be hard to read since it doesn't 'look like C'.

License: Commercial. When you link against the NAG library, people you distribute your program to must also buy the lbrary. There is special pricing for academic institutions (if you're at a university, it may already be installed).

12.5.4 GSL

The GNU Scientific Library (GSL) is a numerical library for C and C++ programmers. It provides a wide range of mathematical routines such as random number generators, special functions, and least-squares fitting. There are over 1000 functions in total. More information can be found at `http://www.gnu.org/software/gsl`.

A simple example (taken from the GSL documentation) is the following program, which computes the value of the Bessel function $J_0(x)$ for $x = 5$:

```
#include <stdio.h>
#include <gsl/gsl_sf_bessel.h>

int main(void)
{
    double x = 5.0;
    double y = gsl_sf_bessel_J0(x);
    printf("J0(%g) = %.18e\n", x, y);
    return 0;
}
```

The output is shown below, and should be correct to double-precision accuracy:

```
J0(5) = -1.775967713143382920e-01
```

License: GPL (*not* LGPL).

12.5.5 FFTW

The FFTW library (the name stands for 'Fastest Fourier Transform in the West') is an example of a dedicated special-purpose library. Developed at MIT, it focuses entirely on performing discrete Fourier transforms. It can handle transforms of arbitrary size and multiple dimensions, and handles parallel computation on systems which have multiple CPUs.

The interesting thing about this library is that it is not optimized for one particular CPU or platform. Instead, it contains a large amount of different implementations, each using special features of specific CPUs (such as SSE, SSE2, 3DNow!, AltiVec), and it selects *at runtime* which algorithm works best for your particular system and the dimensions of your FFT. Before doing the actual FFT calculation, you construct a so-called *plan*, which is optimal for your system, data size, and layout. This plan is constructed simply by trying different algorithms and measuring which is fastest. Each calculation later performed using this plan will use the optimal version of the algorithm. There are also several options when planning; you can do a 'quick guess' which examines a few algorithms which the library thinks are most likely to be fast, or perform an exhaustive search (which can take hours). Obviously, this latter approach is only sensible if you are then going to calculate *many* FFTs, otherwise the few milliseconds saved don't weigh up against the time spent optimizing.

More information can be found on `http://www.fftw.org`.

License: GPL (*not* LGPL), although MIT sells exceptions to the GPL for use in commercial (closed-source) products.

12.5.6 Processor-Specific Libraries

Many processors have special features to help in 'number crunching'. Some of these features are inspired by 'super computer' architectures of the past, such as instructions that operate on whole vectors at a time. This is called SIMD: Single Instruction, Multiple Data. In most modern general-purpose CPUs, these instructions are limited to vectors of four components. This is because when dealing with graphics, both colors and spatial vectors can be represented well in arrays of four values (x, y, z, and 'spare', or 'red', 'green', 'blue', and 'transparency'). In a typical modern system, the CPU is spending a lot of time doing graphical calculations (especially when running games, which is perhaps the strongest market driver for new computer system development) so it made sense to optimize the CPU for those, although the current trend is to offload more and more of these types of calculations to the video card.

Of course, this power can also be used for scientific calculations, and many processor vendors offer mathematical libraries which use the special features provided by their

CPUs. Intel's MKL and AMD's ACML have already been mentioned, but there are more. An example is Intel's IPP (Integrated Performance Primitives).

Some compilers have so-called *intrinsics*, which are C-style functions which provide access to the low-level CPU features. For example, a 128-bit, 4-value vector may be available as an ordinary C datatype (perhaps called something like `__m128`), and when calling special functions with these (like `_mm_mul_ps()` or somesuch), the compiler emits code using the special features of the CPU.

Also, compilers are getting better at detecting opportunities for optimizations using these features, and modern compilers can recognize that code like

```
for (i = 0; i < 4; i++)
    a[i] += b[i];
```

is simply adding two vectors `a` and `b`. In that case, they may substitute this loop for a single 'vector addition' operator. This is called *auto-vectorization*.

12.6 Miscellaneous

There are many, many more libraries available for all kinds of purposes. There are libraries for data compression (zlib for instance; see `http://www.zlib.net`), there are several for reading and writing specific file formats, notably for image files: JPEG (`http://www.ijg.org`), TIFF (`http://www.remotesensing.org/libtiff`), and PNG (`http://www.libpng.org`). Usually, these libraries come with good documentation and example code (the ones mentioned are also all Open Source).

Note that sometimes, these libraries may already be present on (and used by!) your operating system. If not, chances are that using these libraries you get better support for all the 'exotic features' of a certain file format than which the 'native' operating system supports.

For highly optimized JPEG encoding/decoding, you can also check out the commercial libraries by Pegasus Imaging (`http://www.pegasusimaging.com`).

12.7 Synopsis

C has the concept of *libraries*, which are collections of functions which a program can *link* against. There is *static* linking, where the code from the library (or usually just from the functions you call) is included in the program, and *dynamic* linking in which the operating system loads the corresponding libraries into memory when the program is run, and 'patches' the program so that calls to the library execute the correct code.

With dynamic libraries (also called *shared* libraries), the code for the library can be modified independently from the calling code in the program. This means care needs to be taken to ensure *backwards compatibility* so that programs linked against an older version of the library continue to work when the library is updated. One way of ensuring this is with proper *versioning*.

When using third-party libraries, it is important to understand the *license* they are published under. Most licenses impose some kind of restriction on what you are allowed to do with them, especially regarding redistribution (even if they are 'free').

The Standard C library offers a lot of functionality in the areas of math, input/output, string handling, date and time, and generic functions such as sorting and searching and dynamic memory management.

Operating systems usually offer functionality which is available to C programs using libraries provided with them (or installed separately).

There are many libraries offering numerical functionality. They contain algorithms covering linear equations, matrices and vectors, curves, polynomials, special functions, root-finding, series and integrals, transforms (such as FFT), etc. Using the right library can tremendously speed up the development of scientific programs, as well as increase performance.

As was mentioned in §12.4.1, there are languages (such as Java and the .NET languages) which are inherently cross-platform (some would say 'inherently *single* platform', because it only runs on that specific virtual machine).

Most processor vendors offer specific libraries which are tailored to their CPUs.

12.8 Other Languages

There is no other language which has so many libraries available as C does. Since C has become the 'lingua franca' of computer science, even is a library isn't *written* in C, it usually offers *access* from C (and vice versa: many other languages offer a way of interfacing to a C-style library).

It is interesting to note that unlike the C standard library, the C++ standard library *does* make 'performance guarantees': the order of a certain algorithm is part of its specification (*e.g.*, 'this function has $\mathcal{O}(N \log N)$ complexity').

Some other languages (most notably the more modern ones) have very elaborate standard librares. Sometimes these include functionality for building graphical user interfaces. Very few make performance guarantees or contain much functionality for numerical processing. Often, most functionality is focused on 'business software' such as accessing databases or parsing XML files.

12.9 Questions and Exercises

12.1 If your library doesn't have a `strcasecmp` or `stricmp` function for case-insensitive string comparison (see page 257), or you don't want to rely on non-standard functionality, you can write your own. Implement one using the character conversion functions from `ctype.h` (see §12.3.3).

12.2 Check how the `time_t` type is defined in `time.h`, and how it is used to represent a time value on your platform. Given that the type has a finite range, calculate when it 'runs out'. On many UNIX systems, it is a 32 bit integer value which counts seconds since Januart 1st, 1970. Calculate when we are 'out of seconds'. At this time, we can expect a similar situation as with the 'Y2K problem'.

12.3 To get a feeling for the amount of libraries already present on your own system, look into your `/lib` directory for `.so` files (if you're on a UNIX system) or your `C:\WINDOWS\system32` directory for `.DLL` files (if you're on a Windows machine). Or perform a system-wide search for `*.so` or `*.DLL` – how many shared libraries are there on your system? Note that these results are usually skewed because some plugins used by only a single application are also `.so` or `.DLL` files.

12.4 Write a program which inverts a 4×4 matrix. Note that not every matrix has an inverse. Check the result by verifying that the product of a matrix and its inverse yields the identity matrix. For the calculations, you can pick a library – CLAPACK, GSL, or (if present on your system) NAG.

12.5 Combine the knowledge from this chapter and from chapter 11 to write a program which converts image files from JPEG to TIFF or PNG and vice versa.

12.2 Write down the [illegible] type described in [illegible] requires a [illegible] on your platform [illegible] Write [illegible] to [illegible] 32 bit integer [illegible] [illegible] compile and [illegible]

12.3 Do a search for [illegible] on your system [illegible] your [illegible] directory [illegible] your CLASSPATH [illegible] Or [illegible] search [illegible] Note that these results [illegible] because some [illegible] application [illegible] on your [illegible]

12.4 Write [illegible] find where [illegible] parameter [illegible] has an inverse [illegible] the result [illegible] product [illegible] library [illegible] installed on your system.

12.5 Using the [illegible] available from this chapter [illegible] write a program which converts image files [illegible] or PNG [illegible]

Bibliography

[1] James D. Foley, Andries van Dam, Steven K. Feiner, and John F. Hughes. *Computer Graphics: Principles and Practices in C*. Addison-Wesley Professional, 2nd edition, 1995. ISBN 978-0201848403.

[2] Brian W. Kernighan and Dennis M. Ritchie. *The C Programming Language*, 2nd edition, 1988. ISBN 0-13-110362-8 (paperback).

[3] Benoit Mandelbrot. *The Fractal Geometry of Nature*. W. H. Freeman, 1982. ISBN 978-0716711865.

[4] William H. Press, Brian P. Flannery, Saul A. Teukolsky, and William T. Vetterling. *Numerical Recipes in C: The Art of Scientific Computing (Hardcover)*. Cambridge University Press, 2nd edition, 1992. ISBN 978-0521431088.

[5] Andrew S. Tanenbaum. *Structured Computer Organization*. Prentice-Hall, 3rd edition, 1990. ISBN 0-13-852872-1.

[6] Andrew S. Tanenbaum. *Modern Operating Systems*. Prentice-Hall, 2nd edition, 2001. ISBN 0-13-031358-0.

Index

www.ingramcontent.com/pod-product-compliance
Lightning Source LLC
LaVergne TN
LVHW062309060326
832902LV00013B/2127

* 9 7 8 9 0 8 1 2 7 8 8 1 2 *